Isaac Newton's Theory of the Moon's Motion.

Isaac Newton's *Theory of the Moon's Motion* (1702)

With a bibliographical and historical introduction

by

I. Bernard Cohen

DAWSON

ISBN 0 7129 0642 8
First Published by Wm. Dawson & Sons Ltd. 1975.
© Introduction I. Bernard Cohen, 1975

Sole Distributor in the United States
Science History Publications
A division of Neale Watson Academic Publications Inc.
156 Fifth Avenue, New York, New York 10010
Library of Congress catalogue card #75-21565

PHYSICS DEPT.

QB
585
N532

Printed in Great Britain by Unwin Brothers Limited
For Wm. Dawson & Sons Ltd., Cannon House,
Folkestone, Kent, England.

This collection of texts and commentaries on the motion of the moon is respectfully and affectionately dedicated to Prof. Willy Hartner, of the University of Frankfurt, who first introduced me to the history of astronomy, at Harvard in 1935–37.

The illustrations are reproduced by kind permission of the Cambridge University Library. The reproduced copy of Newton's *Theory of the Moon's Motion* (1702) was formerly in the celebrated collection formed by Sir Thomas Phillipps, Bt (1792–1872); it is currently in the possession of the Babson Institute.

I.B.C.

TABLE OF CONTENTS

I. Introduction (by I. B. Cohen)

   Page
   1. Preliminaries 1
   2. The Editions, Translations, and Reprints of Newton's 'Theory of the Moon' 6
   3. Newton's Original Text: English or Latin? .. 12
   4. The English Versions: Was Halley Responsible for the English Edition of 1702? 25
   5. The Astronomical and Practical Significance of Newton's Theory of the Moon 38
   6. The *Theory of the Moon's Motion* (1702) and the *Principia* 55
   7. A Supplemental Enquiry: Was Edmond Halley the Editor of the *Miscellanea Curiosa*? .. 81
   8. Newton's Corrections to the Text of the *Theory of the Moon's Motion* (1702) 87

II. *A New and Most Accurate Theory of the Moon's Motion*... Written by That Incomparable Mathematician Mr. Isaac Newton (London, 1702) .. 89
   To the Reader (p. iii) 91
   The Famous Mr. *Isaac Newton's* THEORY OF THE MOON 97

III. 'Lunæ Theoria Newtoniana': an extract (pp. 332–336) from David Gregory's *Astronomiae Physicae & Geometricae Elementa* (Oxford, 1702) 121

IV. 'Sir Isaac Newton's Theory of the Moon': an extract (vol. 2, pp. 562–571) from David Gregory's *The Elements of Astronomy, Physical and Geometrical* (London, 1715) 129

		Page
V.	'Sir Isaac Newton's Theory of the Moon': Lect. 30–31 (pp. 344–368) of William Whiston's *Astronomical Lectures, Read in the Publick Schools at Cambridge* (London, 1728), containing Newton's tract, presented with Whiston's 'perpetual Explication of the Author's Text'	143

I. INTRODUCTION

1. Preliminaries

Isaac Newton's *Theory of the Moon's Motion* (1702),[1] reproduced below in facsimile, in both the first English and Latin printings, contains a set of procedural rules to be used in making tables of the moon's position. This short tract was composed approximately a decade after Newton had written and published his great *Principia* (1687), but before he had undertaken the final revisions for the second edition (1713). The English version of Newton's 'Theory of the Moon', published as an independent pamphlet in *Theory of the Moon's Motion* (1702), is extremely rare; fewer than a dozen copies have been located,[2] and this work has hardly ever been referred to by scholars.

Newton's treatment of the moon's motion is of special note because his new celestial dynamics, based on gravitation, had enabled him to derive certain 'inequalities' in the moon's motion from a general set of physical causes (*i.e.*, from dynamical considerations). The earlier astronomers had generally used purely empirical procedures based on geometric models. As we shall see below, the unsigned preface to this booklet is of considerable interest, not least because it emphasizes how important it was to the art of navigation to have accurate lunar tables. In this example, there is exhibited the practical or social context of theoretical physical science in the age of Newton.

Newton's *Theory of the Moon's Motion* (1702) was published in the eighteenth century in 4 editions or printings in Latin and 13 in English (in three distinct versions), and so may prove to be one of his compositions most often reprinted in the first Newtonian century. And yet this work has hardly ever been discussed (or even referred to) in the literature concerning Newton or the history of astronomy;[3] and—so far as I know—the particular edition reprinted herewith has been mentioned in print but once,[4] and then without even an indication of the existence of the prefatory essay, 'To the Reader'.

In the following Introduction, I shall first discuss the various printings and translations of Newton's essay, and then turn to two related questions: Did Newton write this essay in English or in Latin? and, Who was the author of the first published English translation (reprinted in facsimile below) and the preface to it? An ancillary question (who was the original editor of the *Miscellanea Curiosa*?) is related to the possibility that Edmond Halley may have been responsible for this first translation of Newton's essay and that, accordingly, he may also have written the preface. It will be seen that there are no unchallengeable answers to these questions. An additional mystery is why Newton chose to permit the document to be published at this particular time and why he allowed it to appear in an astronomical textbook written by David Gregory.

At the end of this Introduction, I shall discuss very briefly the significance of Newton's studies of the moon's motion—from the viewpoint both of Newton's scientific achievement and of the progress of astronomy in general. This will enable the reader to evaluate Newton's *Theory of the Moon*'s *Motion* (1702) as a stage in the unfolding of Newtonian astronomical science. As supplements to the pamphlet, there are three further facsimiles: the text as first printed (in Latin) in David Gregory's textbook of astronomy of 1702, the second English translation of Newton's essay (from the English version of Gregory's book, 1715), and the English translation of the first presentation of this essay with a running commentary (from William Whiston's lectures on astronomy, published in English in 1715).

This assemblage of texts makes available the Latin version in which Newton's essay was first printed, as well as all three of the different translations based upon it. Thus today's reader has conveniently available to him all four versions in which Newton's rules for making lunar tables were read, studied, and used in the eighteenth century. Although it seems almost certain that Newton wrote the essay in English, his own English text was published for the first time only a few years ago; the others were based on the Latin text printed by Gregory, which appears to have been translated from Newton's English into Latin, possibly by Gregory himself.

In what follows, I shall regularly refer to Newton's essay as published in English in the 1702 pamphlet as *Theory of the Moon's Motion* (1702), and I shall refer to the essay itself (without respect to any particular version—Latin or English—or any publication of it), as 'Theory of the Moon'. Thus there should be an unambiguous differentiation of three distinct entities: the essay, its first publication in English, and Newton's theory of the moon in general.[5] The expression 'theory of the moon' (or 'lunar theory') was used in Newton's day somewhat differently from the way it is in ours. At that time, as Baily pointed out, the word 'theory', in astronomy, 'denoted rules or formulæ for constructing diagrams and *tables* that would represent the celestial motions and observations with accuracy'.[6] This usage may be contrasted to 'theory' as 'the development of the *causes* of phenomena, which is its rigid and more modern acceptation.' Newton himself used the word 'theory' in relation to the moon's motion in both senses: as a set of rules for the actual computation of tables of lunar motion, and as the derivation of the 'inequalities' (or apparent irregularities) of the moon's motion from the fundamental principles of dynamics and gravitation. The second sense of 'theory', a derivation of the moon's motion from a causal physical theory, was created by Newton in the *Principia*.[7] Nevertheless, in the context of the essay reprinted in facsimile below, the expression 'theory of the moon' signifies primarily the set of rules (however derived) for computing tables.

Notes

1. As shall be explained below, at the end of this § 1, I have tried to avoid confusion by keeping distinct three similar entities.

(1) *Theory of the Moon's Motion* (1702) = the pamphlet published in 1702, and reproduced below in facsimile, containing an essay (in English) by Newton and an introduction by an unidentified author (and, possibly, editor or translator).

(2) 'Theory of the Moon' = Newton's essay, first published in Latin in David Gregory's astronomical textbook in 1702 and in English in the same year in *Theory of the Moon's Motion* (1702), but without reference to any particular printing or version, whether Latin or English.

(3) theory of the moon = either Newton's rules for determining the moon's position, as given in *Theory of the Moon's Motion* (1702) or elsewhere, or Newton's mathematical treatment of the moon's motion, whether in the *Principia* or any draft version of part of that work or any other manuscript or printed document.

2. Copies may be found in the British Museum; the University Library, Cambridge (England); the Library of Trinity College (Cambridge); the Whipple Museum of the History of Science (Cambridge); the Grace K. Babson Collection, Babson Institute (Mass., U.S.A.). Another copy, from Augustus De Morgan's collection (and previously owned by Francis Baily), is in the library of University College, London. The present facsimile has been made from a copy formerly in Sir Thomas Phillipps's Collection. There are, no doubt, copies in other collections, both public and private. I have not found Newton's copy, which may have been bound up in a volume of 'Tracts Mathematical' or have been part of the 'above one hundred weight of pamphlets and Wast Books' which were in Newton's Library; see R. de Villamil: *Newton: the Man* (London, [1931]), pp. 54, 62, 99. Mr Craig Waff, of the Johns Hopkins University, informs me that there is a copy of this pamphlet in the National Library of Scotland.

3. This essay is discussed at length in the *Supplement* (London, 1837) to Francis Baily's *An Account of the Revd. John Flamsteed* (London, 1835), 'printed by order of the Lords Commissioners of the Admiralty'; a photo-reprint was issued by Dawsons of Pall Mall, London, 1966. In the above-mentioned *Supplement*, Baily reprinted the Latin text of Newton's essay, and referred to various other printings in Latin and in English. But he did not even mention the pamphlet reproduced here, of which he had a copy in his own library (see n. 2 *supra*; and n. 1, § 4, *infra*). Possibly, he acquired this copy after he had published the *Supplement*; or, he might merely have forgotten the existence of

this pamphlet, even though it was in his own collection. The pamphlet does not contain the date of its acquisition by Baily.

4. Not surprisingly (to the *cognoscenti*), the single exception proves to be J. Edleston. In his edition of *Correspondence of Sir Isaac Newton and Professor Cotes* (London, 1850), 'Synoptical View of Newton's Life', p. xxxvi and n. 143, he wrote: '1702 (About June) his "Lunæ Theoria" published in Gregory's Astronomy', and added that this work 'appeared in English, separately, the following August, also in Harris's *Lexicon Technicum*, 1704, (a work to which Newton was a subscriber), and, with a few corrections by Newton in the table of *Errata*, in the *Miscellanea Curiosa*, 1705, (this is the date of the 1st ed., not 1708 as stated by Mr Baily in his *Supplement to Flamsteed's History*, p. 688), with the title of "The Famous Mr Isaac Newton's Theory of the Moon."' On Edleston's further comments, especially in relation to Newton's controversy with Flamsteed, see § 6 of this Introduction, n. 22.

5. See footnote 1 *supra*.

6. *Supplement* (see n. 3 *supra*), p. 690.

7. Prior to Newton, Kepler and Descartes had both dealt with the moon's motion in terms of a causal physical theory, but they were not as successful as Newton, since they did not have the inverse square law of universal gravitation, nor Newton's laws of force and motion.

2. The Editions, Translations, and Reprints of Newton's 'Theory of the Moon'

Newton's 'Theory of the Moon' was first published, in Latin, in David Gregory's *Astronomiæ Physicæ & Geometricæ Elementa* (Oxford, 1702: pp. 332–336), and was in fact one of the significant novelties of that treatise.[1] Newton's essay was reprinted in Latin a few years later, in William Whiston's *Prælectiones Astronomicæ* (Cambridge, 1707: pp. 309–327); and again in the second edition of Gregory's *Astronomiæ . . . Elementa* (Geneva, 1726: vol. 2, pp. 505–513). A somewhat altered or edited version was published by Samuel Horsley in vol. 3 of his edition of Newton's *Opera* (London, 1782: pp. 245–250); the differences between Horsley's text and Gregory's are discussed in § 3 below. In the nineteenth century (as mentioned above, § 1, note 3) this text was printed once again in Latin in Francis Baily's *Supplement to the Account of the Revd. John Flamsteed* (London, 1837: pp. 735–742). In the twentieth century, there have been photo-reprints of Horsley's edition of Newton's *Opera* (Stuttgart-Bad Cannstatt: Friedrich Frommann Verlag, 1964) and of Baily's *Flamsteed* and its *Supplement* (London: Dawsons of Pall Mall, 1966); and a photo-reprint of Gregory's *Astronomiæ . . . Elementa* has been announced (by Georg Olms of Hildesheim).

There have been, furthermore, no fewer than 16 printings or editions of three distinct translations of Newton's essay. The first was the pamphlet, Newton's *Theory of the Moon's Motion* (1702),[2] reproduced here in facsimile, issued with an important introduction that has never been reprinted, and that—as mentioned earlier—has never been discussed or even mentioned in bibliographies or biographies of Newton, or accounts of Newtonian astronomy.[3] A slightly altered or edited version of this first translation appeared within two years, under the heading 'Moon', in the first volume of John Harris's *Lexicon Technicum* (London, 1704), but without the introductory 'To the Reader'; and a year later the first of the three volumes of the *Miscellanea Curiosa* (London, 1705); vol. 1, pp. 270–281) contained a reprint of Newton's *Theory of the Moon* (1702) just as it had been published by Harris. Two further editions of the *Miscellanea* appeared (London, 1708; London, 1726), with Newton's essay;[4] and there were three more editions of vol. 1 of Harris's *Lexicon* (London, 1708; London, 1716; London, 1725), plus a fifth edition (London, 1736), in which the original separate alphabetical sequences of vol. 1 and vol. 2 were combined into a single continuous alphabet running through two volumes, Newton's essay remaining in the *M*'s, under the heading 'Moon',[5] which now appears in the second volume.

A second and wholly distinct English translation appeared in the two-volume edition in English of Gregory's astronomical textbook, entitled *The Elements of Astronomy, Physical and Geometrical* (London, 1715; vol. 2, pp. 563–571),[6] issued again in a revised version made by Edmund Stone, with a new title, *The Elements of Physical and Geometrical Astronomy* (London, 1726; vol. 2, pp. 563–571). The third English version of Newton's essay appears in the translation of Whiston's *Prælectiones*, published as *Astronomical Lectures* (London, 1715; pp. 344–368); reissued as 'The Second Edition corrected' (London, 1728). The revised edition of Gregory's *Elements of Astronomy* (1726) and of Whiston's *Astronomical Lectures* (1728) have both just been reissued in photo-reprint editions (New York and London: Johnson Reprint Corporation, 1972); and the first edition of Harris's *Lexicon Technicum* (London, vol. 1, 1704; vol. 2, 1710) has appeared in a photo-reprint in the same series, 'The Sources of Science' (New York and London: Johnson Reprint Corporation, 1966).

These editions, translations, and reprints may be listed as follows:

LATIN	ENGLISH
*Gregory's *Elementa* (1702)	**Theory of the Moon* (1702)
Whiston's *Prælectiones* (1707)	Harris's *Lexicon* (vol. 1, 1704)
Gregory's *Elementa*[2] (1726)	*Miscellanea* (1705)

LATIN	ENGLISH
Horsley's Newton (1782)	*Miscellanea*[2] (1708)
Baily's *Flamsteed* (1837)	Harris's *Lexicon*[2] (vol. 1, 1708)
Horsley's Newton[2] (1964)	*Gregory's *Elements* (1715)
Baily's *Flamsteed*[2] (1966)	*Whiston's *Lectures* (1715)
Gregory's *Elementa*[3] (1972 [?]: announced)	Harris's *Lexicon*[3] (vol. 1, 1716)
	Harris's *Lexicon*[4] (vol. 1, 1725)
	Gregory's *Elements*[2] (1726)
	Miscellanea[3] (1726)
	Whiston's *Lectures*[2] (1728)
	Harris's *Lexicon*[5] (vols. 1–2, 1736)
	Harris's *Lexicon*[6] (1966)
	Gregory's *Elements*[3] (1972)
	Whiston's *Lectures*[3] (1972)

In the above list an asterisk means that all or part of this work (as much as relates to Newton's essay) is reprinted in facsimile below, following this Introduction.

The above-mentioned editions or printings include all of those which I have encountered. There is one possible further edition, which I have found listed in a bibliography; but I have not included it in the foregoing list, because I have not been able to locate a single copy of it. If Newton's essay did ever appear as listed, this would be an even rarer work than the *Theory of the Moon's Motion* (1702); but I assume it to be a 'ghost'. The reference in question occurs in Jérôme de la Lande [Lalande]: *Bibliographie astronomique* (Paris, An XI=1803), and reads as follows[7] (p. 345):

> 1702. *Londini*—4°. Nova et accurata motuum lunarium theoria, ab Isaaco Newtono anglicè conscripta, et latinè reddita a Davide GREGORIO, M.D. et astronomiæ professore Saviliano.

According to Lalande's preface, he had received many English entries from 'M. Hornsby' of Oxford, *i.e.*, the English astronomer, Thomas Hornsby (1733–1810), Savilian Professor and later Sedleian Professor, and librarian of the Radcliffe Observatory. If, indeed, it was Hornsby who sent this entry to Lalande, then possibly a reference to what appears to be a ghost edition may be found among his papers and correspondence.[8] The title itself is intriguing, because it contains information that cannot be substantiated from any other printed or manuscript source with which I am acquainted: namely, that Newton had originally written his 'theory' in English and that Gregory had then put it into Latin. As shall be seen in § 3 below, all the evidence I have been able to assemble points to the correctness of Lalande's entry, with respect to English as the language in which Newton had composed the 'Theory of the Moon' ('ab Isaaco Newtono anglicè conscripta'). Since it also appears likely that Gregory had translated the 'Theory of the Moon' into Latin for his textbook of astronomy, we are led to wonder how Lalande came by this information ('et Latinè reddita à Davide Gregorio').

The bibliographical entry itself could have resulted from a conflation of the Gregory textbook and the Newton pamphlet, since both are quartos and both are dated 1702. But Lalande gives the work as bearing a London imprint, which is true of the pamphlet in English but not of the textbook in Latin. Furthermore, Lalande's title gives yet further information not available in the pamphlet, Newton's *Theory of the Moon's Motion* (1702), as that Gregory was an M.D. and Savilian Professor of Astronomy. Lalande's entry contains a somewhat weakened form of the English title, in the rendering of 'A New and Most Accurate Theory . . .' as a mere 'Nova et accurata . . . theoria . . .'; accordingly, there may be reason to doubt that Lalande (or some informant) had only translated the English title into Latin, with some additions. If Hornsby was the transmitter of the title to Lalande, we can only wonder where he got it. Born in 1733, Hornsby became an astronomer long after the death of both Gregory and Newton, and so could not have obtained any information at first hand from either of these men, concerning the essay having been written by Newton in English and then translated into Latin.[9]

Notes

1. Gregory's book is noteworthy for its full-length treatment of Kepler's astronomy and for its presentation of Newtonian dynamical astronomy. This work contains two contributions written by Newton. But, whereas the theory of the moon was published by Gregory under Newton's name (see the facsimile below), the other contribution was put into Gregory's preface—thus as if it were from his own pen rather than Newton's. In this document, Newton gave evidence for an ancient tradition of great scientific learning, and declared the great antiquity of much of modern science, even the inverse-square law. See J. E. McGuire and P. M. Rattansi: 'Newton and the Pipes of Pan', *Notes and Records of the Royal Society of London*, 1966, vol. 21, pp. 108–143.

2. I am assuming that Newton's *Theory of the Moon's Motion* (1702) is not the original text of Newton's written in English, but rather an English version based on the Latin text published by Gregory in his *Elementa* (1702). Furthermore, as shall be seen in § 3 *infra*, the Latin text published by Gregory in 1702 appears to have been a translation (very likely by Gregory himself) of an essay written by Newton in English.

3. Possibly, the general lack of references to this booklet may be owing to its rarity (see n. 2, § 1 *supra*). But we may well be puzzled by the paucity of references to Newton's essay by modern scholars, who easily could have encountered it in either Latin or English. But see notes 3 & 4, § 1 *supra*.

4. For the differences among the printings of the *Theory of the Moon* in the editions of Harris's *Lexicon* and of the *Miscellanea*, see § 3 *infra*.

5. I have not personally examined the third and fourth editions, which are not available at Harvard or the British Museum. Since there is no apparent major alteration in the printed text of the 'Theory of the Moon' from the second edition (1708) to the fifth (1736), I presume that there was also no change made in the third and fourth editions.

6. This translation is often said—on what evidence I do not know—to have been made by J. T. Desaguliers.

7. No work with this title appears in the catalogue of the *Bibliothèque Nationale* (Paris), or of the Pulkovo Observatory. I have been unable to find a copy in Harvard, the Babson Collection, the British Museum, the Bodleian, the University Library (Cambridge), or the libraries of Trinity College and King's College in Cambridge. Furthermore, I have not been able to find this title in booksellers' catalogues, notably Sotheran's. Should any reader encounter an actual copy of this work (or any other reference to its existence), I should be most grateful for information concerning it.

8. Since Hornsby was librarian of the Radcliffe Observatory, the catalogues or other lists of that institution (should they exist) might contain a clue. The library was dispersed a number of years ago.

The most recent article on Hornsby, by J. D. North (in the *Dictionary of Scientific Biography*, vol. 6, pp. 511–512), calls attention to a collection of Hornsby manuscripts in the Museum of the History of Science, Oxford (MSS Radcliffe 1–35, 54, 67, 71–73).

9. There is always a possibility that either Lalande himself or some informant had found the title in a list of books—perhaps even a publisher's announcement of a proposed Latin edition that may never have been published. If Newton's *Theory of the Moon's Motion* (1702) had warranted separate printing as a pamphlet in English, why would it not have occurred to some enterprising publisher to issue (in a separate printing) the Latin text on which the *Theory of the Moon's Motion* (1702) had been based?

3. Newton's Original Text: English or Latin?

In studying the circumstances concerning Newton's 'Theory of the Moon', we are handicapped by the fact that this essay is not mentioned in the available correspondence between Gregory and Newton.[1] No Latin manuscript of this essay has ever turned up among the available papers of Newton or of Gregory. We are limited, therefore, to indirect evidence—until (or unless) some positive documentation may be found. In particular we have no way of knowing whether the Latin text in David Gregory's book was written by Newton in Latin as published or was translated into Latin (presumably by Gregory himself) from an English manuscript.[2]

There are some indications that favour the position that Gregory had translated an English manuscript. First of all, there exists just such a manuscript in Newton's hand presently in the Portsmouth Collection, University Library, Cambridge, in § 10 of MS Add. 3966: entitled 'A Theory of the Moon'. A fair copy of this manuscript, in the Library of the Royal Society, London, is written out in the hand of David Gregory; it is dated 27 February 1699/1700 and bears the same title as Newton's own version, 'A Theory of the Moon'. There appears to be little doubt that Gregory's transcript was actually made from Newton's manuscript, since the two texts agree in every respect, but for such 'minor' details as spelling and punctuation; additionally, only Gregory's copy has a date.[3] In the second paragraph (corresponding to pages 11–12 of the booklet), Newton had at first used the word 'aphelium' and only later decided to substitute 'apogee'; this decision must have occurred after Gregory had made his transcript, since the latter contains the word 'aphelium'. In the Latin version published by Gregory, however, this term occurs throughout the second paragraph in the corrected form 'Apogæi Solis' and 'Apogæi Lunæ' or 'Apogæi Lunaris' (and, accordingly, is rendered in English by 'the Sun's Apogæum', 'the Sun's Apogee', or 'his Apogee', and by 'the Moon's Apogee', 'the

Lunar Apogee', or 'her Apogee', on pp. 11–12 of Newton's *Theory of the Moon's Motion* (1702), as reproduced below. A comparison of the MS English essay[4] and Gregory's printed Latin version shows without question that one is a translation of the other—there being a near-perfect agreement between the two, paragraph by paragraph.

Moreover, there exists among Newton's papers an earlier draft in English of the essay copied out by David Gregory. It is entitled by Newton 'The Theory of ye Moon', which he later changed, when he made the final version, to 'A Theory of the Moon'. This preliminary version occupies all four sides of a single folded sheet. The first page is partially cancelled,[5] so that the actual text copied by Newton begins only on the verso of the first page, as follows:

> The mean motions of y^e Sun & Moon from y^e Vernal
> Equinox I put as follows vizt anno 1680 upon y^e last day of
> December at Noon stilo veteri the mean motion of the Sun
> 9^s. 20^g. $34'$. $46''$.of his Aphelium. . . .

In the final version, Newton has added an introductory paragraph, but only after copying out the text as written above. This additional short paragraph, dealing with the position of the 'Observatory at Greenwich',[6] is accordingly squeezed into the space between the title and the above paragraph, and is written in a very compressed handwriting. In a number of examples, it is easy to see that the partial MS is indeed prior to the more complete version, since alterations made in the text of it are embodied in the final copy. For instance, in the early draft, we may see that Newton wrote:

> There are annual equations of ye motions of ye Sun &
> Moon & of her Apoge and Node.

Then, he put a caret before 'There are . . . ' and inserted:

> These motions are tempered wth several inequalities &
> first

In the final copy, all of this is combined into a single opening sentence,[7] now revised so as to read:

> The mean motions above described are affected w^{th}

several inequalities. And first there are annual equations of the mean motions of y^e Sun & Moon & of her Apoge & Nodes.

Again, in the next paragraph 'There is another Equation of the Moons mean motion . . .'[8]), the words 'in other positions of the Sun it is reciprocally proportional to y^e cube of the distance of y^e Sun from $y^e \ominus$' are an insert, as is the 'backwards' in a later part of this paragraph ('. . . in the passage of the Apoge backwards from y^e Quadratures to y^e Syzygies'). In the final MS version, both of these insertions are incorporated in the text proper. There is a sufficient number of other such examples to remove any possible doubt that the version from which Gregory made his copy was a final version, based by Newton on an earlier English draft.[9] Such evidence is convincing that Newton wrote this essay in English, but it does not tell us whether or not Newton himself later composed the Latin version, which Gregory printed in his book in 1702.

In an attempt to resolve the question as to whether or not Newton may have produced a Latin version as well as an English text, we may compare the style of the published Latin text with known examples of Newton's composition in Latin.

Gregory's final paragraph begins with the words, 'Telluris Atmosphæra', and later mentions 'Telluris Umbram'. Some years later, Newton wrote out this same paragraph for the second edition of the *Principia* (1713), with some major changes. Here (Scholium to Prop. 35, Book III) we find Newton using the phrases, 'atmosphæra terræ' and 'umbram terræ'. Not only did Newton use the noun 'terra' rather than 'tellus'; he also put the adjective after the noun, rather than before it. Lest it be thought that the word-order and choice of noun may have been due to editorial intervention (either by Roger Cotes or Richard Bentley), let me point out that Newton's MS drafts of this scholium correspond in this regard to the printed version. This test indicates that the style of the Latin text published by Gregory differs from Newton's usage.[10]

In the opening sentence, we may also discern a difference

between Newton's Latin style and that of the text published by Gregory in his textbook of astronomy. The Latin version given by Gregory begins:

> Observatorium *Grenovicense* occidentalius est *Parisiensi* 2.^{gr} 19′. . . .

I believe Newton would have written this somewhat differently, presumably in the form:

> Observatorium *Grenovicense* est occidentalius Observatorio *Parisiensi* 2^{gr}. 19′. . . .

This would be more in conformity with the style of a Latin document written by Newton, and dated Apr. 1700, on this same topic, where we find:

> Uraniburgum est orientalius observatorio Regio *Parisiensi* 00^h. 42′. 10″ & hoc Observatorium est orientalius *Grenovicensi* 00^h. 09′. 15. . . .
>
> [Uraniburg is east of the Royal Observatory at Paris by 00^h. 42′. 10″ and this observatory is east of that of Greenwich by 00^h 09′. 15″ . . .][11]

It would seem, therefore, that the Latin sentence published by Gregory is a translation, by someone other than Newton himself, of Newton's sentence in English (from the English draft of this essay), reading:

> The observatory at Greenwich is more westward then ye observatory at Paris by 2^{gr}. 19′, then Dantzick by 18^{gr}. 48′, then Uraniburg by 12^{gr}. 51′. 30″, then Rome by [12° 30′].

The difference in style between Gregory's text and Newton's authentic Latin writings may be seen also by comparing a Latin sentence in the essay with Newton's drafts and final version of the Scholium to Prop. 35, Book III, of the *Principia*. In Gregory's Latin text, there is a sentence immediately following the one concerning the Greenwich Observatory (discussed in the preceding paragraph), which begins as follows:

> Solis & Lunæ Motus medios ab Æquinoctio verno in Meridiano *Grenovicensi* pono sequentes. . . .

In the Scholium, in the second edition of the *Principia* (1713), however, we find:

Et ... motus medios solis & lunæ ad tempus meridianum in observatorio regio *Grenovicensi*, ... sequentes adhibebit. ...

Not merely is the phrasing greatly different, but the order of Newton's words 'motus medios solis & lunæ' (in the Schollium) differs from Gregory's 'Solis & Lunæ Motus medios'.

There are, however, a few sentences in Gregory's Latin text that are very close indeed to sentences written out by Newton for the revised Scholium to Prop. 35, Book III, and eventually printed in the second edition of the *Principia*. I give three of the most striking examples, in each case italicizing the words that are identical in both versions, and that appear in the same order.

[Gregory, p. 333, par. 2]
Alia est Æquatio *Motûs medii* Lunæ, *pendens à situ Apogæi* Lunaris respectu Solis, *quæ maxima est cum Apogæum Lunæ versatur in Octante cum Sole, &* nulla cum llud ad Syzygias *vel* Quadraturas pervenerit.

[MS Add. 3966, fol. 65]
Et hinc oritur *alia æquatio motus medii* Lunaris *pendens a situ apogæi* Lunæ ad Solem, *quæ* quidem *maxima est cum apogæum Lunæ versatur in Octante cum Sole*, et nulla cum illud ad quadraturas *vel* syzygias pervenit.[12]

Apart from the context, which in one case demands a supplementary statement ('Alia est ...') and in the other a conclusion from previous theory ('Et hinc oritur ...'), the only differences are Newton's use of 'lunaris' for 'Lunæ' in the phrase, 'motus medii lunaris' and of 'lunæ' for 'Lunaris' in 'apogæi lunæ', the substitution of 'ad solem' for 'respectu Solis', the extra word 'quidem', the change in order of 'quadraturas 'and 'syzygias', and the shift in tense from 'pervenerit' to 'pervenit'. But the identity of the remaining words themselves, and their order in the sentence, could hardly be owing to chance.

Next, consider two successive sentences, corresponding to a pair that appears on page 17 of Newton's *Theory of the Moon's Motion* (1702), the first completing the upper paragraph and the second beginning the lower paragraph on that page:

[Gregory, p. 333, par. 3]	[MS Add. 3966, fol. 65]
Additur hæc *Motui Lunæ dum Nodi transeunt à Solis Syzygiis ad* ejusdem *Quadraturas*; *& subducitur in eorum transitu à Quadraturis ad Syzygias*.	*Additur* vero medio *motui Lunæ dum nodi transeunt a Solis syzygiis ad* proximas *quadraturas & subducitur in eorum transitu a quadraturis ad syzygias*. . . .
[Gregory, p. 333, par. 4]	[MS Add. 3966, fol. 65v]
A Solis Loco vero aufer Motum medium Apogæi Lunæ *æquatum*, ut supra est ostensum; *residuum erit Argumentum Annuum* dicti *Apogæi*.	*A Solis loco vero aufer motum medium apogæi* sic *æquatum* et *residuum erit argumentum annuum apogæi*.

The near-identity of these parallel sentences, shown by italicizing the words and phrases that are identical in both, may seem to constitute overwhelming evidence that Newton must have given Gregory his own Latin version. Yet we must be cautious in hastily drawing any such conclusion. For, when Newton came to write up the new or revised Scholium to Prop. 35, Book III, he would most likely have had Gregory's printed Latin text available to him, and could easily have incorporated some of the actual sentences of Gregory's version into his own Scholium. The closeness of these few odd sentences may thus not indicate an identity in style so much as a borrowing by Newton of a 'Newtonian' text that had actually been put into Latin by Gregory from an essay written by Newton in English.[13]

The textual analysis of the essay published by Gregory (and by Horsley) does not, of course, enable us to make an unambiguous decision as to whether Newton or Gregory was indeed responsible for the version published in Latin. On this subject, Gregory said, in introducing Newton's essay (page 332 of his *Astronomiæ* . . . *Elementa*, reproduced in facsimile below):

. . . In hac Calculi forma, quam ipsis Auctoris verbis expressam Astronomis sistimus . . .

In the English translation (1715, vol. 2, p. 562), this reads:

. . . In this Calculation, which we give in the Words of the Author. . . .

The Latin original of Gregory's introduction contains a stronger statement concerning 'the Words of the Author' than is found in this English version, since the latter does not give the full stress of the words 'ipsis . . . verbis expressam'. No doubt, almost all readers would have assumed that Gregory meant that he was giving them the 'very words of the author' in the language in which the author had written them, and that the Newtonian text was not a paraphrase or a translation.

In the preface 'To the Reader' (p. vi of the *Theory of the Moon's Motion* (1702) printed below), the writer-translator says of Newton's essay that:

> *I thought it would be of good service to our Nation to give it an English Dress . . . . I hope we have a great many Persons in England that have Skill and Patience enough to calculate a Planet's Place, who yet it may be don't well enough understand the Latin Tongue to make themselves Masters of this Theory in the Author's own Words.*

The translator thus had no doubt that he was rendering into English a composition in Latin by Newton. On p. 9, he repeats Gregory's statement, declaring that Newton's 'Theory . . . is now publish'd in Dr. *Gregory*'s Astronomy, in Mr. *Newton*'s own Words.'

The title-page of the pamphlet would accordingly be interpreted by any reader so as to give the same conclusion. The essay is there described in these words:

Written by
That Incomparable Mathematician
Mr. Isaac Newton,
And Published in *Latin* by
Mr. David Gregory
in his *Excellent Astronomy*.

While this 'blurb' does not say unambiguously that the essay was 'written in *Latin*' by Newton and then 'Published' by Gregory 'in his *Excellent Astronomy*', I doubt whether anyone would have interpreted it to imply that Gregory had published his own Latin version of Newton's essay 'in his *Excellent Astronomy*', especially in view of the more explicit statement in the foreword 'To the Reader'.

Since we have no evidence on which to make a positive identification of the translator (and author of the foreword) of *Theory of the Moon's Motion* (1702), we cannot evaluate the basis on which he made so positive an assertion that Newton had written his essay in Latin. Thus we cannot tell whether his statement merely was derived from a reasonable interpretation of Gregory's introductory paragraph in his astronomy textbook, or was based on any other independent sources of information. Since Gregory could honestly have intended his reader to understand only that Newton's *very own words* had been translated, or that Newton's theory was not being restated or described or summarized or paraphrased, the evidence of the pamphlet is just as ambiguous on this point as that of Gregory's own introduction.

In the eighteenth century, a somewhat variant Latin version was published in vol. 3 of Horsley's edition of Newton's *Opera* (London, 1782). Horsley, however, gives no indication of his source; and I do not believe we have any grounds on which to suppose that he did anything other than reprint Gregory's version, but with some minor changes. Horsley's presentation differs from Gregory's in three major respects. First, there is a different division into paragraphs, to which numbers from 1 to 17 are assigned (three paragraphs being unnumbered). Of the 17 numbered paragraphs, 15 are printed by Horsley with postils.[14] But we have no basis for saying whether or not these postils were composed by Horsley himself. In the absence of any other text (MS or printed) with these postils, we may assume that they were editorial additions introduced by Horsley.

These conclusions concerning Newton's 'Theory of the Moon' may be summarized as follows. All the available evidence very strongly favours the supposition that Newton's essay was originally written as the English composition, of which he retained the final version (now preserved in the Portsmouth Collection in the University Library, Cambridge), and which was copied out by Gregory, whose transcript survives among his papers in the Library of the Royal Society. Furthermore, each of the three English versions published in the eighteenth century is a direct translation from the Latin text of Newton's essay published in 1702 in Gregory's textbook of astronomy; these translations include Newton's *Theory of the Moon's Motion* (1702), essentially reprinted in Harris's *Lexicon Technicum* and the *Miscellanea Curiosa*, and the respective English translations of Gregory's textbook and of Whiston's *Prælectiones*. Finally, it seems to be the case that the Latin version of Newton's 'Theory of the Moon', published by Gregory in 1702, and reprinted with editorial additions and emendations in Horsley's edition of Newton's *Opera*, was a translation from English into Latin, possibly made by Gregory himself.

Notes

1. See vol. 4 of Newton's *Correspondence*, ed. by J. F. Scott (Cambridge, 1967), which covers the years in question; supplemented by W. G. Hiscock (ed.): *David Gregory, Isaac Newton and their Circle: Extracts from David Gregory's Memoranda 1677–1708* (Oxford, 1937).

I have found no information concerning this essay in the two volumes of Gregory MSS in the library of the Royal Society, nor in the MS copies of Gregory's *Notæ in Isaaci Newtoni Principia Philosophiæ* for which see I. B. Cohen: *Introduction to Newton's 'Principia'* (Cambridge, 1971), ch. VII, §§ 12–13, and W. P. D. Wightman: 'David Gregory's Commentary on Newton's "Principia"', *Nature*, 1957, vol. 179, pp. 393–394.

There may, of course, be a reference to this topic among Gregory's miscellaneous MSS, most of which are to be found in the Edinburgh University Library; a preliminary search disclosed no such information.

2. As we shall see below, there is some evidence that suggests that Newton wrote out this essay in English. The question of whether he himself made a translation into Latin, which Gregory published, or whether Gregory made the translation, is more difficult to answer, but we shall see that there are aspects of style that seem to suggest a translator other than Newton himself.

INTRODUCTION 21

3. In a memorandum dated 21 May 1701, Gregory stated his wish: '2. To gett the æquatiuncula in the Theory of the Moon of Mr Newton . . . as Mr Hally wrote to me.' If we could find this letter from Halley to Gregory, we might know more about Newton's *Theory of the Moon's Motion* (1710). See Newton's *Correspondence*, vol. 4, pp. 354–355.

4. A transcript of this English essay is given in vol. 4 of Newton's *Correspondence*, pp. 322–327. But, although the text is alleged to be that in U.L.C. MS Add. 3966, it lacks the title which there appears in Newton's hand. Furthermore, the date '27 February 1699/1700' does not appear in Newton's MS, but only in the copy Gregory made of it; presumably, then, this is the date on which Gregory made his copy, and not the date of composition. In the notes accompanying this document, no mention is made of an earlier draft in this same MS Add. 3966, a few folios away from the final version (but the same § 10). An editorial reference is given to 'a Latin translation in Horsley', but without any mention of the Latin translation in Gregory's astronomy textbook, nor of the several English versions.

5. In cancelled or rejected portions, Newton wrote:

'I reccon y^e tropical year to consist of 365 days 5 hours 48 minutes 57 seconds & by consequence the Sun's mean motion in four solar years or 1461 days to be 48 signes 1 minute 47 seconds & $55\frac{1}{2}$ thirds. . . .

'I reccon y^e mean motions of y^e Sun & Moon to be as follows Anno 1700 upon y^e last day of December at noon stilo veteri the mean motion of y^e Sun from y^e vernal equinox 9^s. 20^{gr}. $43'$. $50''$, of y^e Suns Aphelium 3^s. 7^{gr}. $4.3'$. $20''$, of y^e Moon. . . .

'And that in 20 Julian years or 7305 days y^e mean motion. . . .'

6. This corresponds to the second paragraph on p. 10 of the *Theory of the Moon's Motion* (1702).

7. This sentence corresponds to a combination of the second and third paragraphs on p. 13 of the *Theory of the Moon's Motion* (1702).

8. This corresponds to the second paragraph on p. 15 of the *Theory of the Moon's Motion* (1702).

9. This prior draft is not complete as it stands, ending at a point corresponding to the middle of the second paragraph on p. 21 of the *Theory of the Moon's Motion* (1702). An additional single MS page (one side only) corresponds *grosso modo* to some other parts of the final version of Newton's essay. More information concerning this group of MSS is given in § 6 *infra*.

10. In the *Principia*, two forms of 'tellus' occur: 'tellurem' (used once) and 'telluris' (used eleven times); but there are 391 occurrences of the forms of 'terra'.

11. These MSS may all be found in U.L.C. MS Add. 3966.

12. Here, and in the next two examples, I have contrasted Gregory's text with fol. 65 of MS 3966, rather than the second edition of the *Principia*, to eliminate any possibility that the printed versions may differ in actual expression from the texts written out by Newton himself. In the second and third editions of the *Principia*, these texts are very much altered from the MS drafts. See, e.g., the variant readings for p. 461, lines 22–23 in A. Koyré, I. B. Cohen, Anne Whitman (eds.): *Isaac Newton's Philosophiæ Naturalis Principia Mathematica: the third edition (1726) with variant readings* (2 vols.: Cambridge, 1972).

13. Evidence for Newton's possession of a copy of 'Gregorii Astronomiæ Physicæ &c. Elementa, F. 1702' is given in R. de Villamil: *Newton: the Man* (London [1936]), p. 78.

Newton's own copy survives, and is presently in the Trinity College Library, Cambridge.

14. Horsley's text diverges from Gregory's Latin version only in three very minor details: [i] Horsley uses 'prosthaphæresis' for Gregory's 'Prostaphæresis' (p. 334, par. 2, lin. 4 bott.); [ii] Horsley uses commas for Gregory's parentheses to mark off the phrase 'e contra' (p. 335, par. 2, lin. 3): [iii] Horsley uses a different figure for the one presented by Gregory on p. 334, with the result that angles, arcs, line-segments, and the circle itself are designated in the figure (and referred to in the text) by letters not always identical to those used by Gregory.

The figure used by Horsley appears to have been taken from the Scholium to Prop. 35, Bk III, of Newton's *Principia* (ed. 3: London, 1726). I suppose that Horsley reprinted the Latin text given by Gregory, assuming that it was the original of Newton's, and only changed it superficially with regard to the above-mentioned paragraphing and numbering, adding postils, changing the spelling (and capitalization) of a single word, replacing a pair of parentheses by commas, and introducing a substitute figure (together with a few resulting changes in designation of parts of the figure).

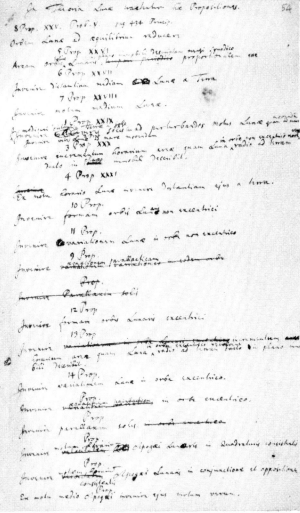

One of Newton's Plans for revising the Portion of Book III of the *Principia*, Relating to the Motion of the Moon.

(The page reference is to the first edition; this plan as not adopted in the seconded edition. This manuscript is in Newton's handwriting.)

MS A^(dd.) 3966, University Library, Cambridge

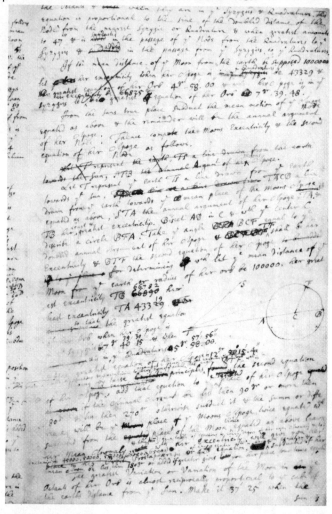

Work Sheet for Newton's Essay in English on the Theory of the Moon's Motion (MS A^{dd.} 3966, University Library, Cambridge. This page is in Newton's handwriting.)

4. The English Versions: Was Halley Responsible for the English Edition of 1702?

In Gregory's *Elements of Astronomy* (London, 1715: vol. II, p. 562), Newton's essay is introduced by a paragraph, translated without significant alteration from the Latin edition (Oxford, 1702: p. 332), as follows:

Scholium

I have thought fit to subjoyn the Theory of the Moon made use of by Sir *Isaac Newton*, by which this incomparable Philosopher has compass'd this extremely difficult Matter, hitherto despair'd of by Astronomers; namely, by Calculation to define the Moon's Place even out of the Syzygies, nay, in the Quadratures themselves so nicely agreeable to its Place in the Heavens (as he has experienc'd it by several of the Moon's Places observ'd by the ingenious Mr. *Flamstead*) as to differ from it (when the difference is the greatest) scarce above two Minutes in her Syzygies, or above three in her Quadratures; but commonly so little that it may well enough be reckon'd only as a Defect of the Observation. In this Calculation, which we give in the Words of the Author, he does not wholly mention all the Inequalities, whose Causes are above explain'd, nor those which are as yet only suspected; but omitting those which he knew wou'd take off one another, and others of less Moment, he only confines those to Æquations and Tables, that have the greatest Force and produce the most sensible Effects.

Following Newton's essay, Gregory has placed a paragraph of his own (p. 571), reading:

If several Places of the Moon nicely observ'd (chiefly about the Quadratures) be compar'd with the Places of it calculated for the same Time in the Theory above; it will appear whether or no there are any sensible Æquations wanting to make it more perfect.

In the Latin edition (p. 336), of which a portion is printed below in facsimile, this paragraph of Gregory's is clearly demarcated from the preceding paragraphs of Newton's. All lines of type in the extended quotation from Newton begin with inverted commas (or quotation marks), whereas such marks are absent from Gregory's own paragraphs. But in the English versions of Gregory's textbook (London, 1715: vol. 2, pp. 562–563, 571; London, 1726: vol. 2, pp. 562–563, 571) no typographic distinction of this sort differentiates Gregory's two paragraphs. The first, or introductory, paragraph could hardly have been written by Newton, since it refers to him as 'this incomparable Philosopher'; furthermore (as may be seen in the facsimile reproduction of the relevant pages, to be found following this Introduction), Gregory's prefatory paragraph is set off from Newton's opening sentence by a title, reading:

Sir Isaac Newton's *Theory of the Moon.*

But the last paragraph appears as if it were part of Newton's text, which it is not. We shall see below (§ 6) that there exists some confusion on these points, since the only mention of Flamsteed occurs in the introductory paragraph written by Gregory; Newton never refers to Flamsteed by name in the body of his essay, the 'Theory of the Moon'.

In the pamphlet reproduced below, Newton's *Theory of the Moon's Motion* (1702), this final paragraph is introduced (p. 29) by a sentence not found in either the Latin or the English version of Gregory's textbook: 'Thus far the Theory of this Incomparable Mathematician.' This sentence alerts the reader to the fact that the final paragraph is not Newton's, and in context there could be little doubt in any reader's mind that the conclusion is Gregory's. In the pamphlet, *Theory of the Moon's Motion* (1702), furthermore, Gregory's opening phrases have been suppressed to make way for a new introductory paragraph (p. 9), obvious to all as not having been written by Gregory, and reading:

> This Theory hath been long expected by the Lovers of Art, and is now publish'd in Dr. *Gregory*'s Astronomy, in Mr. *Newton*'s own Words.

The second introductory paragraph then more or less agrees with the first part of Gregory's introduction. But it will be seen that the editor has understandingly suppressed the second part discussing Newton's failure to 'wholly mention all the Inequalities, whose Causes are above explain'd' (that is, explained by Gregory in the preceding part of the chapter on the moon's motion; in the astronomy textbook), 'nor those which are as yet only suspected'. I shall discuss below some further differences between the opening paragraph in *Theory of the Moon's Motion* (1702) and that in the English translation of Gregory's textbook.

The following sets of extracts will suffice to show the differences among the four English versions, and to prove that the three translations are to a real degree independent of one another.

[Newton's MS (U.L.C. MS Add. 3996, §10, fol. 74)]

The mean motions above described are affected wth several inequalities. And first there are annual equations of the mean motions of ye Sun & Moon & of her Apoge & Nodes. The annual equation of the Suns motion arises from ye excentricity of his Orb. . . .

There is another equation of the Moon's mean motion wch depends upon the position of her Apoge to ye Sun & is greatest when the Moon's Apoge is in the Octants of the Moon's Orb & vanishes when it is in the syzygies & quadratures.

[Newton's *Theory of the Moon's Motion* (1702), pp. 13, 15]

These mean Motions of the Luminaries are affected with various Inequalities: Of which,

1. There are the Annual Equations of the aforesaid mean Motions of the Sun and Moon, and of the Apogee and Node of the Moon.

The Annual Equation of the mean Motion of the Sun depends on the Eccentricity of the Earth's Orbit round the Sun . . .

There is also an *Equation of the Moon's mean Motion* depending on the Situation of her Apogee in respect of the Sun; which is *greatest* when the Moon's Apogee is in an Octant with the Sun, and is nothing at all when it is in the Quadratures or Syzygys.

[Gregory's *Elements of Astronomy* (1715), pp. 564, 565]

The mean motions of the Luminaries abovementioned have several Inequalities.

First there are the Annual Æquations of the said mean motions of the Sun and Moon, and of the Apogæum and Node of the Moon. The Annual Æquation of the mean motion of the Sun depends upon the Excentricity of the Earth's Orbit about the Sun. . . .

There is another *Æquation of the Moon's mean Motion*, depending upon the situation of the Moon's Apogæum in respect of the Sun, which is the greatest when the Apogæum of the Moon is in the *Octant* (or at half Right-angles) with the Sun, and none at all when it is come to the Syzygies or Quadratures.

[Whiston's *Astronomical Lectures* (1715), pp. 348, 352]

The Mean Motions of the Luminaries above suppos'd, are affected with many Inequalities: And first, there are the Annual Equations of the said Mean Motions of the Sun and Moon, and of the Apogee and Nodes of the Moon: The Annual Equation of the Mean Motion of the Sun depends upon the Eccentricity of the Orbit of the Earth about the Sun. . . .

There is another Equation of the Mean Motion of the

Moon, depending upon the Situation of the Lunar Apogee with respect to the Sun; which is greatest of all when the Moon is an Octant with the Sun; and none at all when it comes to the Conjunctions; [Conjunctions I say in the Plural, for under that Word I here include the Opposition also] and Quadratures.

The three translators not only produced independent versions, but clearly worked without access to Newton's MS essay. It may seem more than a little astonishing that three separate translations should have been made of an essay of Newton's which was available in Newton's own words in English! A closer examination, however, discloses that at least two of the translators most likely would not have known of the existence of Newton's original draft. The translation of Gregory's textbook appeared in 1715, seven years after Gregory's death; there was no reason why the translator (possibly J. T. Desaguliers) should have known that Gregory had had in his possession an English version of this essay, written by Newton. Whiston's volume of *Astronomical Lectures* was published in English without the name of the translator (who may be presumed to have been Whiston himself); but there was no reason for author or translator to have supposed that the text taken from Gregory's *Astronomiæ . . . Elementa* had not been printed in Newton's original Latin version.

Our interest concentrates, therefore, on the translation given in the *Theory of the Moon's Motion* (1702). I have mentioned above that this text, without the prefatory essay, was reprinted in the *Miscellanea Curiosa* (vol. 1, London, 1705), in a version differing from the *Theory of the Moon's Motion* (1702) only in the most minor details, chiefly printing style. It was the very fact of this reprint, coupled with the supposition that Edmond Halley had been the editor of the *Miscellanea*, which led Augustus De Morgan to assume that Halley had been responsible for the original English version, *Theory of the Moon's Motion* (1702), and the prefatory essay.

De Morgan's copy of the 1702 pamphlet, at present in the library of University College, London (see note 1, § 1 *supra*), contains a number of annotations. On the title-page, he has written his name, 'A. De Morgan', and the following note:

> This copy belonged to Francis Baily and was the one used by him in his Flamsteed.

But De Morgan was mistaken with regard to Baily's having used this booklet when he wrote and published either his *Flamsteed* or the *Supplement*, since Baily does not refer to this version in listing the publications of Newton's essay.[1] De Morgan himself recognized the error in the note he had written on the title-page about Baily's having used this copy 'in his Flamsteed', but he did not then correct this first comment or cancel it. Rather, he proceeded to give the correct information in a long MS annotation which occupies the otherwise blank verso side of the title-page. De Morgan's note reads *in extenso* as follows:

> In the interval between the first and second editions of the Principia, Newton communicated his amended lunar theory to David Gregory, by whom it was published in the work on Astronomy, Oxford, 1702, 4to. In the same year, this theory was translated into English, and separately published in the tract of which this is a copy. This English publication, and its preface, are no doubt due to Halley, who inserted this same English in the *Miscellanea Curiosa*, published in 1708.
>
> Baily, to whom this copy belonged, seems not to have known of any English earlier than 1708, when he published his work on Flamsteed (see Flamsteed, Suppl. p. 688). This first edition is excessively rare: I never saw any other copy.
> *October 9, 1858.*   A. DE MORGAN
>
> This is the first work in which Newton suppressed his acknowledgments of Flamsteed's assistance. Flamsteed's name only appears in the paragraph by Gregory. So little was this tract known, that when Baily said Flamsteed was

not mentioned in the *Theoria Lunæ*, many thought the *lunar theory* in the first edition of the Principia was spoken of, or else went to Horsley's Newton, and set down Baily as quite wrong.

Following the title of the prefatory 'To the Reader'—occurring on p. iii of Newton's *Theory of the Moon's Motion* (1702)—De Morgan wrote: 'by Halley, no doubt'.[2]

The supposition that Halley was the author of the translation, and of the preface 'To the Reader', is an appealing one. Halley was, in 1702, one of a handful of English mathematical astronomers who could fully comprehend and appreciate the value of Newton's theory of the moon. Additionally, he was very much interested in the possible applications of that theory to the problem of determining the longitude. He might very well have been responsible for the view expressed on p. iv of the preface 'To the Reader' of Newton's *Theory of the Moon's Motion* (1702):

> ... *could her [i.e., the moon's] Place be but truly calculated, the Longitudes of Places would be found every where at Land with great Facility, and might be nearly guess'd at Sea without the help of a Telescope, which cannot there be used.*

In his essay on the 'true Theory of the Tides', Halley had expressed these very sentiments concerning 'the Inequalities' in the motion of the moon and Newton's work on this problem:

> ... And tho' by reason of the great Complication of the Problem, he has not yet been able to make it purely Geometrical, 'tis to be hoped, that in some farther Essay he may surmount the difficulty: and having perfected the Theory of the Moon, the long desired discovery of the Longitude (which at Sea is only practicable this way) may at length be brought to light, ...[3]

Furthermore, when Halley reported to Newton, on 5 July 1687, that the printing of the *Principia* had been completed, he urged the illustrious author ('after you shall have a little diverted your self with other studies') to take up anew 'the perfection of the Lunar Theory'. It would be worthy of Newton's 'attempt', he said, since a successful Lunar Theory 'will be of prodigious use in Navigation'.[4] There can be no doubt, therefore, that the sentiments expressed in the preface to this booklet accord perfectly with what we know to have been Halley's concerns.

The style of the foreword is very much like Halley's—straightforward and factual. Furthermore, the title-page refers to Newton as 'that Incomparable Mathematician', a striking phrase that is immediately reminiscent of the opening line of Halley's review of the *Principia*; there he calls Newton 'This incomparable Author'.[5]

While the foregoing information makes very plausible the suggestion of a possible role for Halley in the production of the *Theory of the Moon's Motion* (1702), there is no real evidence that I have been able to find to substantiate the case for Halley. I know of no letter from or to Halley or Newton, referring to the translator or editor of the 1702 tract.

Let me turn now to the version of Newton's 'Theory of the Moon' printed in the *Miscellanea Curiosa*, reserving until later (§ 7 *infra*) the question of whether Halley was the actual editor or compiler of this collection. As mentioned earlier, this version (*Miscellanea*, 1705: vol. 1, p. 270) is substantially the same as the Newtonian portion of the *Theory of the Moon's Motion* (1702); but the preface ('To the Reader') of the tract has been omitted altogether. In the *Miscellanea*, as in the tract (p. 9), the essay is headed, '*The Famous Mr.* Isaac Newton's *Theory of the Moon.*' Save for the opening paragraph, the text found in the *Miscellanea* differs from that given in the *Theory of the Moon's Motion* (1702) only in such minor printing details as punctuation, capitalization, spelling, etc. (For instance, the *Miscellanea* has '12 deg. 51 min. 30 seconds' where the booklet reads '12°. 51′. 30″.') The concluding paragraph (p. 29 of the booklet) is the same in both printings, although the final reference to Gregory and the word 'FINIS' do not appear in the *Miscellanea*. The word 'FINIS', however appropriate at the end of a small book, would have been wholly out of place in an article in the *Miscellanea*; the information concerning the pages in Gregory's book (given at the end of *Theory of the Moon's Motion* (1702), p. 29) was not needed in the *Miscellanea*, since the page numbers were there given in a new introductory paragraph:

[*Miscellanea* (1705: vol. 1, p. 270)]

This Theory which hath been long expected by all the true Lovers of *Astronomy*, was communicated from Mr. *Newton*, to Dr. *Gregory Astron*. Professor at *Oxford*, and by him published in his *Astron. Elem. Philos.* and *Geomet.* p. 336. From whence, as it was lately translated into *English*, I thought fit to insert it here.

[*Theory of the Moon's Motion* (1702)]

This Theory hath been long expected by the Lovers of Art, and is now publish'd in Dr. *Gregory's* Astronomy, in Mr. *Newton's* own Words.

Finally, in the *Miscellanea*, as in the *Theory of the Moon's Motion* (1702), no typographical distinction is made between the editorial introductory and concluding paragraphs and Newton's text.

In the front matter of the *Miscellanea*, on the last page of the 'A' gathering, following 'The Table [of Contents]', there is printed a set of 'Corrections made by Mr. *Isaac Newton* to the *Theory of the Moon*.' Their presence on this page indicates that Newton had communicated his emendations, directly or indirectly, only after the main body of the *Miscellanea* had been printed off, since otherwise the changes would have been incorporated in the text itself. We have no way of telling whether Newton had seen the printed-off pages of the *Miscellanea* and had then made some corrections, or whether Newton had entered corrections in either a copy of the pamphlet or of the earlier reprint in John Harris's *Lexicon Technicum* (London, 1704).[6] I have listed these corrections at the end of this Introduction, so that each reader may correct the text of the *Theory of the Moon's Motion* (1702) in accordance with Newton's instructions of 1705.

The final sentence of the introductory paragraph, just quoted from the *Miscellanea*, ends with the clause 'I thought fit to insert it here.' At first encounter, this would seem to be a statement by the editor of the *Miscellanea*, explaining how he had come to reprint Newton's 'Theory of the Moon' in the collection.[7] But this introductory paragraph was not composed for the *Miscellanea* by its editor, but was taken directly from the earlier reprint of Newton's essay by John Harris. Accordingly, it cannot be taken as evidence of Halley's participation in the publication of the first English translation, the *Theory of the Moon's Motion* (1702).

The argument that the 'English publication [of 1702], and its preface are no doubt due to Halley' was advanced by De Morgan without any other ground than that Halley 'inserted this same English in the *Miscellanea Curiosa*'. But if the *Miscellanea* contains the very same text as Harris's *Lexicon*, then it could be more properly argued that the translation (or at least its first publication, and its preface) was due to Harris and not to Halley at all. De Morgan would have been misled on this point by his lack of awareness of the earlier reprint of Newton's *Theory of the Moon's Motion* (1702), in John Harris's *Lexicon Technicum* (London, 1704: in the article MOON). As a matter of fact, the *Miscellanea* took its text directly from Harris's *Lexicon* and not from the booklet itself. The proof of this statement may be found in the fact that some of the details of printing style (in which the *Miscellanea* differs from the 1702 pamphlet) occur in the *Lexicon*.[8] Furthermore, not only are the opening or introductory paragraphs of the essay in the *Lexicon* and the *Miscellanea* identical in almost every detail (and thus distinctly different from the pamphlet), but the *Miscellanea* even continues such a misprint as *Urainburgh* from the *Lexicon*, whereas this word occurs correctly in the pamphlet (p. 10, par. 2, line 3) as *Uraniburgh*.

Even though it was John Harris (and not Halley) who first declared, 'I thought fit to Insert it here', Harris does *not* say that he had been the translator. In his general preface, he merely says that he has included in his *Lexicon* 'in particular, the Incomparable Mr. *Isaac Newton*'s *Theory of the Moon*: and a very large Account of *Comets* from the same Author'. While Harris's English summary of Newton on the comets, taken from Book III of the *Principia*, shows him to have had the competence to translate Newton's essay on the moon and to write a preface to Newton's essay, there is no real evidence to support the hypothesis that he may actually have done so.

Unless positive evidence may be forthcoming, therefore, I do not believe there are any real grounds for assigning the authorship of the translation or preface to Halley.[9] Halley's candidature is not only weak to begin with, but it rests on the supposition that Halley was the editor or compiler of the *Miscellanea*, which—while likely—is not absolutely certain.[10] Halley's candidacy need not be given serious consideration until or unless new evidence turns up to support it; his name was put forth in ignorance of the fact that the reprint of Newton's essay in the *Miscellanea* was not made from the *Theory of the Moon's Motion* (1702), but rather from an earlier reprint in Harris's *Lexicon*. Our lack of knowledge as to the identity of the translator prevents us from even attempting to solve another mystery: why, in 1702, a translation was made from Gregory's Latin version, when Newton's original essay could have been readily available[11] from either Newton himself or from Gregory.

Notes

1. It must therefore follow that Baily either had acquired his copy only after he had published the *Supplement* to his life of Flamsteed, or had forgotten all about the publication in English of Newton's *Theory of the Moon's Motion* (1702) of which there was a copy in his personal library.

2. Other than the foregoing MS comments, here published for the first time, I do not know of any published note by De Morgan on this

work of Newton's, nor have I found a reference to it in his published essays on Newton. A complete annotated list of De Morgan's notes and essays, with a comprehensive index, is a great desideratum.

A 'List of Writings' is given on pp. 401–415 of Sophia Elizabeth De Morgan's *Memoir of Augustus De Morgan* (London: Longmans, Green, and Co., 1882), but it is admittedly incomplete.

3. This essay was published as a pamphlet, and then reprinted in *Phil. Trans.*, 1697, No. 226, pp. 445–457. A facsimile reprint appears in I. B. Cohen (ed.): *Isaac Newton's Papers & Letters on Natural Philosophy* (Cambridge, Mass., 1958), pp. 412 sqq. See, further, A. N. L. Munby: 'The Distribution of the First Edition of Newton's *Principia*', *Notes and Records of the Royal Society*, 1952, vol. 10, p. 33.

4. Newton's *Correspondence*, vol. 2 (Cambridge, 1960), p. 482.

5. *Phil. Trans.*, 1687, No. 186, p. 291; reprinted in *Newton's Papers & Letters* (see n. 3 *supra*).

The similarity in these phrases was emphasized by Sir Edward Bullard, with whom I corresponded concerning Halley as the possible author of this preface and translator of *Theory of the Moon's Motion* (1702). He not only found the general style to be 'characteristic' of Halley, but drew my attention in particular to a stylistic similarity between the preface and Halley's letter (concerning Newton's work) to John Wallis, of 9 April 1687; see MacPike's edition of Halley's letters (n. 29, § 6 *infra*), p. 80. On the other hand, the preface omits the 'e' in such words as 'look'd' (p. iii, l. 5 of the preface), whereas Halley tended to use the 'e' in his MSS and in his published articles in the *Phil. Trans.* We have no way of telling whether this particular form may have resulted from the printer's peculiarities rather than the author's.

6. Harris's reprint is discussed almost immediately below.

7. According to the title of the first edition, the *Miscellanea Curiosa* was 'A collection of some of the Principal Phænomena in Nature, Accounted for by the Greatest Philosophers of this Age', printed 'Together with several Discourses read before the Royal Society'. But in the second edition, the title was altered, the phrase 'Together with several Discourses' being replaced by 'Being the Most Valuable Discourses, Read and Delivered to the Royal Society' which would make the presence of Newton's essay anomalous, since it had neither been 'Read' nor 'Delivered to the Royal Society'.

8. E.g., the spelling out of 'seconds' and 'minutes' or the use of 'deg.' and 'sec.' and 'min.' in place of the symbols. But Harris's text contains many more capitalized words than does the *Miscellanea*.

9. The foregoing discussion arises from the fact that Halley's name has been put forth in this connection by De Morgan; but there

is currently no real evidence whatever to support De Morgan's guess, which—in any event—was put forth in ignorance of the fact that the version of the 'Theory of the Moon' in the *Miscellanea* had previously been published in Harris's *Lexicon Technicum*.

10. See the separate discussion of this question, below, in § 7.

11. The mystery is compounded by the fact that Harris reprinted the English version, *Theory of the Moon's Motion* (1702), as if it had been newly translated for the *Lexicon*; and that the reprint in the *Miscellanea* was similarly presented, as if it had been freshly translated rather than being a reprint from the *Lexicon*. Another curious aspect of the affair is that Newton must have known of the Harris reprint, since he was a subscriber to the *Lexicon*; and he must have known of the *Miscellanea* reprint, since he provided corrections and owned the volumes. And yet he allowed the continued reprinting of a translation of a translation, without ever putting into press his own original version.

5. The Astronomical and Practical Significance of Newton's Theory of the Moon

The foregoing sections of this Introduction have dealt with bibliographical questions respecting the publication of Newton's *Theory of the Moon's Motion* (1702), together with aspects of the composition, translation, and subsequent printings of this essay. In the present section, the general astronomical significance of Newton's analysis of the motion of the moon will be examined in relation to the theory of gravitation and the problem of the longitude; while, in the concluding section (§ 6), some technical aspects of the *Theory of the Moon's Motion* (1702) will be discussed in relation to the successive editions of the *Principia*.[1]

A revolutionary feature of Newton's *Principia* was the partial demonstration that the moon's motion might be deduced mathematically from a system of physical causes producing their effects according to exact laws. In this sense, Newton's study of the moon's motion was different from most other work on this topic, from remotest antiquity to 1687 (when the *Principia* was published). Two exceptions are Johannes Kepler (in Book VI of the *Epitome*) and René Descartes (in his *Principia* and in correspondence with Mersenne, published in the 17th century). But generally, before Newton's *Principia*, the study of the moon's motion had been a branch of celestial geometry, in which varieties of models were constructed to fit the data of observation; in these models, circles were combined with other circles, both fixed and variable and both homocentric and excentric.[2]

After the *Principia* had been published, the subject of the moon's motion became a central part of celestial dynamics, in which the major causes for the irregularities (or 'inequalities') of the moon's motion have been sought in the action of gravitational forces, including perturbing forces. The dramatic change in lunar theory brought about by the *Principia* may be discerned in the fact that Newton's gravitational analysis not only led to predictions of magnitudes and periods of the several known 'equations' (or 'inequalities'[3]), but also to their conditions for maximum and minimum values; while Newton's mathematical analysis disclosed, additionally, the existence of certain unsuspected 'equations' that were derived only from theory, but would not at that time have been found from direct observation. We shall see below that in Newton's *Theory of the Moon's Motion* (1702), and in the *Principia* (especially in ed. 2, 1713, and in ed. 3, 1726), both types of inequalities occur in the presentation: the classical inequalities and the new ones that Newton himself had discovered.

The difficulties in any lunar theory, or the prediction of the moon's place,[4] or the complexities of the moon's motion, arise from a number of separate causes. First of all, as Newton showed, in any system of two gravitating bodies, each body moves about the common centre of gravity. Because the sun has so much greater a mass than any planet, this centre of gravity for the earth-sun system is very near the centre of the sun; but in the case of the earth-moon system, the centre of gravity is only about 1000 miles in from the earth's surface along a line from the earth's centre to the moon's centre, thus requiring a significant modification of Kepler's laws (as Newton found). The moon is subject to a major force, the gravitational pull of the earth, and also to a second significant force, the gravitational pull of the sun, which produces perturbations. This perturbing force is not constant, but varies both in magnitude (in relation to the sun's distance from the moon) and in direction (according to the changing position of the sun relative to the moon). Additionally, the earth-moon force is not constant, since the moon's orbit is non-concentric and the earth itself is moving in a non-constant manner, subject to the pull of the moon and the forces of perturbation of other planets. In fine detail (although Newton did not get that far in his analysis), even the perturbing force of Venus must be taken into account.[5] Accordingly, the motion of the moon must be analysed into irregularities in orbital motion that involve a changing orbit that varies in both its shape or magnitude and orientation, and is subject to a variety of motions, including a regression of the line of nodes (or intersection of the plane of the moon's orbit and the ecliptic), a periodic variation in the inclination of the orbit, and a perturbation of the period of revolution.

Basically, as we shall see below (in § 6), the Newtonian method is to begin with the 'two-body problem', in which the moon and earth mutually attract one another according to the law of universal gravitation, a law of the inverse-square of the distance. For this model, Newton shows how Kepler's laws must be modified, and how each of these bodies will move in an orbit about their mutual centre of gravity. Then Newton observes that both the moon and earth will be acted on by the sun's gravitational force. It is only the 'accelerative measure' (or the acceleration) of the gravitational force that matters, not the force itself,[6] and it acts differently on the two bodies according to their respective distances from the sun and their orientation: this difference is the 'disturbing' force (or its 'accelerative' measure), as Newton called it, and it is denoted by us as the perturbing force.[7] Although Newton did not consider the possible perturbations produced in the moon's motion by the planets, he was aware of the mutually perturbing effect of one planet upon another, notably in the case of so massive a planet as Jupiter.[8]

Newton did not ever succeed in producing a complete solution of the problem of the motion of the moon in relation to sun and earth (even neglecting the possible influence of planets on the moon's motion); today we are aware that this 'three-body problem' cannot be solved analytically, and there is no such solution even for the restricted problem of three bodies.[9] It is as true today as it was when Rouse Ball remarked:

> The general problem of the motion of three bodies under their mutual attraction still remains unsolved, and that Newton should have been able, with the limited analysis at

his command, to work it out so far in the case of the moon, is worthy of special notice.[10]

Airy referred to 'Newton's eleventh section' (the portion of Book I of the *Principia* in which Newton uses dynamical principles to develop the inequalities in the motions of the moon) as 'the most valuable chapter that has ever been written on physical science'. Laplace made an unequivocal assertion concerning Newton's applications of the general dynamical principles (Book I, Sec. 11, *Principia*) to the actual observed lunar motion (in Book III): 'je n'hésite point à les regarder comme une des parties les plus profondes de cet admirable ouvrage.'[11]

There is no want of evidence that it was the revolution in the approach to the problems of lunar motion that particularly aroused the admiration of Newton's contemporaries and successors. As the reviewer of the second edition of the *Principia* in the *Acta Eruditorum* (Leipzig, March 1714, p. 140), put it:

> Indeed, the computation made of the lunar motions from their own causes, by using the theory of gravity, the phenomena being in accord, proves the divine force of intellect and the outstanding sagacity of the discoverer.[12]

It must not, however, be thought that the general recognition of Newton's genius, in pioneering a wholly new way of dealing with the moon's motion, necessarily has ever implied a universal agreement with his own results in detail. For instance, one of the outstanding examples of dissatisfaction with Newton's lunar theory arose from the problem of the motion of the lunar apogee. The observed value of the mean motion of the apogee was just about twice as large as the predictions based upon Newtonian theory, a fact which at first led both Clairaut and Euler 'to infer that the inverse-square law of gravitational attraction should be regarded only as an approximation to the truth'.[13]

Great as Newton's personal contribution was, the lunar theory—as presented in the successive editions of his *Principia* and in the *Theory of the Moon's Motion* (1702)—was only in its infancy in 1726, when the last authorized edition of the *Principia* appeared under Newton's direction. The imperfections in that theory, and its limitations, were primarily due to the want of fully adequate analytical tools; yet it must continually be kept in mind that the problem of three bodies cannot be fully solved analytically. The investigations to be made after Newton were of such magnitude and of so high a level of difficulty as to require more than the efforts of a single mathematician: even an Euler or a Laplace or a Poincaré! This quality of the problem of the moon's motion is made manifest by the parade of geniuses of the highest order, as well as mathematicians of more than ordinary genius, who have made such major contributions to our understanding of the motion of the moon: including (following the age of Euler, Clairaut, Tobias Mayer, and d'Alembert), such mighty figures as Lagrange, Laplace, Poisson, Gauss, Jacobi, J. C. Adams, Le Verrier, Sir George H. Darwin, G. W. Hill, C. E. Delaunay, F. Tisserand, Henri Poincaré, E. W. Brown, C. V. L. Charlier, G. D. Birkhoff, Tullio Levi-Civita, Aurel Wintner, Harald Bohr.[14]

Reference has already been made to the practical importance assigned to studies of the moon's motion in Newton's day, and for a century thereafter. Thus to Newton's contemporaries, his research on lunar motion was more than just a witness to his intellectual prowess (in the sense expressed by the reviewer in the *Acta Eruditorum*, 1714) or a proof of the validity and usefulness of the theory of gravity (as Newton said of his own research on lunar theory, in the second edition of his *Principia*, 1713[15]). As the author of the preface to Newton's *Theory of the Moon's Motion* (1702) made explicit, the theoretical determination of the moon's position was a key to finding the longitude at sea. In 1714, the year following the second edition of the *Principia*, Parliament established 'a Publick Reward for such Person or Persons as shall discover the Longitude at Sea'. The prize was to be £10,000 for a means of determining a ship's longitude to within 1 degree of arc, but £15,000 if the method were good to 40' of arc, and £20,000 if to 0.5 degrees of arc or 30'. The members of the Board of Longitude, set up to judge proposals and to give general advice, included such notable astronomers as Edmond Halley, John Flamsteed, and Isaac Newton himself.[16] To see how important this problem was, we may go beyond the actual value of the prize (the maximum award being estimated to be of the order of some $1,000,000 or about £400,000 in today's money) and recall certain naval disasters attributable to a failure in determining longitude. In 1691, four years after the first publication (1687) of the *Principia*, seven British warships were so ignorant of their position in longitude that they took the Deadman for Berry Head and were wrecked near Plymouth. Three years later, in 1694, Admiral Wheeler's fleet, ignorant of its position, sailed head-on into Gibraltar and disaster. In 1707, Sir Cloudisley Shovell's squadron of the Royal Navy, depending on the erroneous opinions of the navigators of different ships, believed themselves in a safe position, only to run aground on rocks off the Scilly Isles, with a loss of 200 lives and four ships of the line.[17]

Many apparently promising methods failed to produce the desired result; they made use of dead reckoning, the variation and dip of a magnetic compass needle, the eclipses of the satellites of Jupiter, sound signals which could be timed, and various astronomical phenomena that either occur too infrequently to serve for determining longitude when needed (e.g., solar eclipses, transits of Mercury and Venus, occultations of various stars by the moon) or are impractical for observation with necessary precision on ship-board (e.g., meridional transits of the moon).[18] The two most promising (and rival) proposals for finding the longitude were the direct use of a marine chronometer and the method of 'lunar distances'. It had become well known during the seventeenth century that the accurate determination of longitude at sea was an astronomical problem, or depends on celestial observations—whether of stars, planets (or their satellites), sun, or moon. The Royal

Observatory was established at Greenwich, as the Royal Warrant (4 March 1675) declared, so that the Astronomer Royal (Flamsteed was the first one) might produce improved tables 'of the Motions of the Heavens and the Places of the Fixed Stars, in order to find out the so much desired Longitude at Sea, for perfecting the Art of Navigation.'

Since terrestrial longitude may be reckoned by the difference between the local time and the known time on a standard meridian (say Greenwich), the longitude at sea can be found by any method giving the difference between the navigator's local time at any instant and the local time at Greenwich at that same instant. The purpose of the marine chronometer was to provide the navigator with an accurate mechanical device so that (with suitable corrections which are a function of the particular clock in use) he might know the Greenwich local time at any instant at which he determined his own local time. In the method of lunar distances, the standard of comparison came from a set of tables rather than a mechanical clock.

The method of lunar distances became the primary system advocated by British astronomers for finding the longitude at sea. While depending primarily on tables of the moon's motion, it also required a well designed and accurately made observational instrument. In practice, the navigator must determine a lunar distance, that is, the angle between the moon and a celestial object, say a star. This observation then requires a number of correction factors: for atmospheric refraction, for the fact that the lines of sight to moon and star are made from the earth's surface and not from the earth's centre, for the possibility that measurements may have been made from the edges of the moon's limb rather than the centre. In fact, rather than in principle, the navigator would first determine his latitude (not too difficult) and then the observed altitude of the moon and of some one of the chosen reference stars. The corrected lunar distance could be compared with the true value of the angular distance of the moon to that star, as given in a printed table as a function of time; the difference determines the longitude. Of course, all of this sounds incredibly simple as theory, but was far from that simple in practise. It may be pointed out, however, that tables of lunar distances became a feature of the *Nautical Almanac* from its inception (1765), and that the method of determining longitude in this manner was standard throughout most of the nineteenth century.[19]

In order to use the method of lunar distances, ship-board navigators had to have accurate and easy-to-use instruments (e.g., the sextant) to measure the apparent altitude or zenith distance of the moon and a star (or the sun, or a planet), plus tables of correction for atmospheric refraction, etc., and reduction tables to convert the result into longitude. But the primary requirement was obviously a correct table of the moon's motion, so as to make possible an accurate series of predicted lunar distances, or true geocentric angular distances from the moon to a designated set of stars (or planets, or the sun). There was required the development of a means of predicting future positions of the moon accurately, such as Newton's breakthrough—the new science of celestial dynamics—promised to yield. Furthermore, a long series of reliable lunar observations was needed, in order to test and to correct the theory.[20] In Newton's day, the extended series of lunar observations made at the Royal Observatory by Flamsteed, and then by his successor Halley, were thus of crucial importance. And we may readily understand why any apparent reluctance (much less downright refusal) by Flamsteed to make his lunar observations available to Newton would have seemed to the latter to have been an issue of national significance and not a mere personal matter, as we shall see below in § 6.

Both Newton and Halley were aware that the first edition of the *Principia* had not dealt with the actual motion of the moon in a manner that would lead to the construction of adequate tables, such as might be useful for navigation. On sending Newton the news that the *Principia* had at last reached the final stages of printing, Halley said (5 July 1687), that he hoped that,

... after you shall have a little diverted your self with other studies, ... you will resume those contemplations, wherin you have had so good success, and attempt the perfection of the Lunar Theory, which will be of prodigious use in Navigation, as well as of profound and subtile speculation.[21]

We know that Newton himself was not satisfied with the practical state of lunar computations at the time of the first edition of the *Principia* (1687). For in that edition, in Book III, in which he applied to 'The System of the World' the theoretical principles developed in Books I and II, a Scholium, following one of the propositions on the moon's motion, concluded:

I don't care to add the computations, however, as being too complicated and encumbered by approximations, and not accurate enough.[22]

Newton was referring here specifically to the immediately preceding computation of apsidal motion.

In the second edition, as we shall see below in § 6, Newton cast out this Scholium for a lengthy new one, giving the history (in part) of his researches on lunar motion and setting forth some 'equations' for correcting the lunar theory, the latter based on the rules contained in his *Theory of the Moon's Motion* (1702), which was intermediate in publication between the first (1687) and second (1713) editions of the *Principia*.

In the light of the foregoing discussion, the reader will surely find it odd that in his own text in the *Theory of the Moon's Motion* (1702), Newton himself makes no reference either to the theoretical grounding of his own research, based on considerations of gravitational celestial mechanics, or to the importance of an accurate theory of the moon's motion for the solution of the problem of finding the longitude at sea. Nor does David Gregory refer either to Newton's deduction of the equations from gravity or to the usefulness of those same equations for navigation, in his textbook of astronomy. As may be seen in Gregory's own comments in the paragraphs at the beginning and end of Newton's essay (printed below, in facsimile, in both Latin and in English), he says only that he is giving the reader 'the Theory of the Moon made use of by Sir *Isaac Newton*', in order to calculate the position of the moon 'even out of the Syzygies, nay, in the Quadratures themselves'. And he adds that the maximum difference between the computed positions and Flamsteed's observations are 'scarce above two Minutes in her Syzygies, or above three in her Quadratures'; and the difference is 'commonly so little that it may well enough be reckon'd only as a Defect of the Observation'. Gregory also says that Newton does not 'wholly mention all the Inequalities, whose Causes are above explain'd' (that is, those explained by Gregory in his prior explicative summary of the *Principia*); nor does Newton mention inequalities which 'are as yet only suspected'. Newton has thus omitted inequalities 'which he knew wou'd take off one another, and others of less Moment', and has produced 'Æquations and Tables' based only on those inequalities 'that have the greatest Force and produce the most sensible Effects.'

In the preface ('To the Reader') to the pamphlet reprinted below, Newton's *Theory of the Moon's Motion* (1702), however, there is a direct reference to Newton's great intellectual achievement in deriving the inequalities in the moon's motion from the theory of gravity. On pp. iv–v, the editor (or translator) states that:

> *This Irregularity of the Moon's Motion depends ... on the Attraction of the Sun, which perturbs the Motion of the Moon (and all other* Satellites *...) and makes her move sometimes faster and sometimes slower in her Orbit; and makes consequently an Alteration in the Figure of that Orbit, as well as of its Inclination to the Plain of the Ecliptick.*

This fact is, indeed,

> *now well known, since Mr.* Newton *hath demonstrated the Law of Universal Gravitation.*

The importance of this discovery of how to calculate the moon's position, furthermore, is that if the moon's position could be

> *truly calculated, the Longitudes of Places would be found every where at Land with great Facility, and might be nearly guess'd at Sea without the help of a Telescope, which cannot there be used.*[23]

Such a point of view may thus be contrasted with Newton's essay itself, which—as just mentioned and as may be seen by examining it in the various versions that are reprinted below—never once mentions any connection between the rules it contains and the physical theory of the moon developed in the *Principia*. So independent is this essay from the mathematico-physical doctrines of the *Principia* that Whiston included it at the end of his lectures on descriptive and computational astronomy, the series that preceded his 'Explication of Sir *Isaac Newton*'s Astronomy and Philosophy', which (as he said in his lecture on 6 December 1703) would be deferred 'until the next Term'.

In short, however Newton may have derived the rules which are set forth in the 'Theory of the Moon', he did not there present them in relation to any mode of derivation; they are merely set forth as a series of rules for computing. The whole essay consists of a set of instructions, one after the other, which (if followed exactly) will supposedly yield the moon's position accurately. Were it not for the author's name, and the nature of the treatment of the moon in his previously published *Principia*, no reader would ever have guessed that these procedural rules might possibly have been derived from the new dynamical principles (as Newton was to allege in the second edition of the *Principia*).[24] And, in fact, Whiston (see his Lecture XXXI and p. 361 of his *Astronomical Lectures*, printed in facsimile below) admits his bewilderment concerning Newton's derivation of the '6th Equation of the Moon'; and he even voiced a suspicion 'that this Equation was rather deduc'd from Mr. *Flamsteed*'s

Observations, than from Sir *Isaac Newton*'s own Argumentation'.

Notes

1. An introduction to Newton's *Theory of the Moon's Motion* (1702) is hardly the place for an extensive conceptual and mathematical analysis of Newton's treatment of the motion of the moon in all three editions of the *Principia*, plus the manner of his further work on this topic as revealed by his many MS essays, notes, and computations. Accordingly, in this brief survey, only enough information will be presented to enable the reader to understand the historical significance of Newton's revolutionary innovations in lunar theory and the relation of the 'Theory of the Moon' to the whole corpus of his research in this area.

2. From antiquity onward, it had been known that the moon's motion could not be represented by a model in which the moon revolves uniformly along a simple circle concentric with the earth, if the positions were to agree even roughly with observations. Astronomers conceived of a regular or 'mean motion' of the moon, on which were superimposed certain 'inequalities' which would produce a more complex model, giving a better agreement with observations. In these models, not only were regular motions in circles combined with yet other motions in circles (homocentric or excentric), but the motion could be either uniform in arc along a circle or uniform in angle about an equant, a point which in ancient astronomy could be distinct from either the earth or the geometric centre. Such epicyclic systems, however, were not designed to produce retrograde and direct motion, as in the case of planets, but to regularize an irregular direct lunar motion.

The Ptolemaic construction (featuring a 'crank' device) had been notable for sacrificing correct apparent distances for desired position in angle; the substitute Copernican construction, long thought to have been one of the distinctive (and original) features of the system of *De Revolutionibus*, was anticipated by (if not, indeed, taken from) the work of ibn al-Shāṭir. Although Kepler introduced a celestial physics based on causes (as the full title of his *Astronomia Nova* (1609) declared), purely geometric models of the moon's motion continued to be constructed throughout the seventeenth century, of which the most notable was Flamsteed's Horroxian lunar theory (for which see n. 19 to § 6 *infra*). Even Newton, in the new Scholium following Prop. 35, Book III, ed. 2 of the *Principia* (1713), discussed below, introduced such a model along with dynamical considerations and deductions from the theory of gravity.

3. In older science the expression 'equation' signified both an error of some sort (or a discrepancy from regularity) and the correcting term (i.e., the correction applied to a quantity). Thus the 'equation of time' is the quantity that expresses the principal difference between local apparent solar time and mean time (or between the longitude of the true sun and of the mean sun); the 'personal equation' is the correction that must be applied to measurements as a result of the individual characteristics of an observer; the 'annual equation' is 'any inequality in the elements of an orbit, or in a position, whose period is one year'; accordingly, an 'equation' may signify both an 'inequality' or a correction to such an 'inequality'.

4. In § 1 *supra* (see especially n. 5 to § 1), it was shown that there were two meanings to 'lunar theory' or 'theory of the moon' in Newton's day. One was a set of procedural rules used in computing tables of the moon's motion or position. The other was a set of physical principles (e.g., gravitational celestial dynamics) from which the motions—and hence the positions—of the moon could be deduced.

5. For a rather heroic job of exposition of the moon's motion, without the use of mathematical equations, see G. B. Airy in *Gravitation: An Elementary Explanation of the Principal Perturbations in the Solar System* (London, 1834). A more technical treatment of the same subject by Airy occurs in his *Mathematical Tracts on the Lunar and Planetary Theories* . . . (ed. 3, Cambridge, 1842). Both works were written with direct references to Newton's *Principia*. A somewhat shorter non-mathematical account of the lunar theory may be found in Sir John F. W. Herschel's *A Treatise on Astronomy* (London, 1833), ch. 11, §§ 553 sqq.

One of the most valuable systematic introductions to classical (Newtonian and post-Newtonian) lunar theory is Hugh Godfray: *An Elementary Treatise on the Lunar Theory, with a brief sketch of the history of the problem before Newton* (ed. 3, London and New York: Macmillan and Co., 1871), and the lengthy historical Introduction to John N. Stockwell's *Theory of the Moon's Motion deduced from the Law of Universal Gravitation* (Philadelphia: Press of J. B. Lippincott & Co., 1881) may be highly recommended.

See, further, Richard Stevenson: *Newton's Lunar Theory, exhibited analytically* (Cambridge, 1834).

Some other causes of inequalities in the motion of the moon are the oblateness of the earth (which, because of its figure, does not attract the moon as if all of its mass were concentrated at its centre), and the motion of the ecliptic; these are admirably presented by Godfray, *op. cit.*, ch. vii, §§ 108–113.

6. In the beginning of the *Principia*, Newton set forth three 'measures'

of centripetal force, which he designated (Defs. 6–7–8) as the 'absolute quantity', the 'accelerative quantity', and the 'motive quantity'. On this subject see I. B. Cohen: 'Isaac Newton's *Principia*, the Scriptures, and the Divine Providence', pp. 523 sqq. of *Essays in Honor of Ernest Nagel: Philosophy, Science, and Method* (New York: St. Martin's Press, 1969), Appendix I: 'New Light on the Form of Definitions I–II & VI–VIII', pp. 537–542.

The 'accelerative quantity' is defined as the 'measure' of a 'centripetal force', 'proportional to the velocity which it generates in a given time.' Any combination of the law of gravitation ($F = G \times mM/r^2$) for the gravitational force of a body of mass M at a distance r, acting on a body of mass m, with the second law of motion ($f = mA$) for the effect on m, yields a law ($A = GM/r^2$) that measures the acceleration A. Thus, as a standard astronomical textbook puts it, the '*accelerations*, and not the *forces*, are what really count. The force of the sun's attraction on the earth is more than eighty times as great as that on the moon, owing to the difference of mass; but the accelerations at the same distances are equal. This use of the term "force" instead of "acceleration" is common in celestial mechanics. It is noteworthy that exact calculations can be made of the motion of a body (for example, a comet) when we have no idea what its mass is and consequently do not know the force of attraction ($f = GmM/r^2$); for the acceleration ($f/m = GM/r^2$) is accurately known.' See Henry Norris Russell, Raymond Smith Dugan, John Quincy Stewart: *Astronomy*, I: The Solar System (Boston: Ginn and Company, 1926), pp. 286–287.

7. It will be seen below (§ 6) that in the fundamental first presentation of perturbations in the *Principia* (Prop. 66 and its 22 corollaries, Book I), Newton is primarily concerned with the motion of the moon, as is made clear by the designation of T ($= Terra$) for the central body about which the secondary planet P circulates, while being perturbed by the force of an outer moving body S ($=Sol$). In the first edition, Newton was less explicit, having S ($=Sol$) as the central body, about which a planet P ($=Planeta$) circulates, while being perturbed by an outer planet Q. Since the corollaries apply primarily and directly to the problems of the motion of the moon, Newton changed the letters from Q-P-S in ed. 1 to S-P-T in ed. 2, presumably at the suggestion of Fatio de Duillier; see I. B. Cohen: *Introduction to Newton's 'Principia'* (cited in note 1 to § 3 *supra*), ch. 7, § 9, pp. 181 sqq.

8. In 1684–1685, he corresponded with Flamsteed on the possible effect of Jupiter on the motion of Saturn; see Newton's *Correspondence*, vol. 2, pp. 406–415.

9. The restricted problem of three bodies is a special case of the three-body problem, in which: 'Two bodies S and J revolve round their

centre of gravity, O, in circular orbits, under the influence of their mutual attraction. A third body P, without mass (i.e., such that it is attracted by S and J, but does not influence their motion), moves in the same plane as S and J; the restricted problem of three bodies is to determine the motion of the body P, which is generally called the *planetoid*.' Quoted from E. T. Whittaker: *A Treatise on the Analytical Dynamics of Particles and Rigid Bodies, With an Introduction to the Problem of Three Bodies* (ed. 3, Cambridge: at the University Press, 1927), p. 353. The best available treatment of the restricted problem is given in Aurel Wintner: *The Analytical Foundations of Celestial Mechanics* (Princeton, New Jersey: Princeton University Press, 1941), ch. 6, 'Introduction to the restricted problem'. According to Wintner, pp. 348 sqq.: 'The restricted problem of three bodies was first considered by Euler in connection with one of his lunar theories. The mathematical and astronomical significance of this model was, however, understood only much later. First, Jacobi observed that the problem is, as a matter of fact, a conservative problem with two degrees of freedom. . . . This conservative formulation of the restricted problem . . . became fundamental, first in Delaunay's elaborate lunar theory, and then . . . during the last quarter of the 19th century. On the one hand, G. W. Hill developed at that time his lunar theory, which . . . as elaborated in its details by E. W. Brown, is to-day the most precise treatment of a problem ever dealt with in celestial mechanics. . . . At the same time, this model aroused the interest of Poincaré, whose mathematical work in dynamics centered about it . . . considered . . . as a prototype of those dynamical problems which have two degrees of freedom and are not "integrable" in the sense in which a problem with a single degree of freedom is. . . . At any rate, almost everything mathematically significant in the progress of general analytical mechanics during the 20th century, and in particular the dynamical work both of Levi-Civita and of [G.D.] Birkhoff, was originally directed towards, when not influenced by, an investigation of the restricted problem of three bodies.' Wintner gives 'historical notes and references' on this topic on pp. 437sqq. of his book.

10. W. W. Rouse Ball: *An Essay on Newton's 'Principia'* (London and New York, 1893), p. 89; this work has been reprinted by the Johnson Reprint Corp., (New York and London, 1972).

11. G. B. Airy: *Gravitation: An Elementary Explanation of the Principal Perturbations in the Solar System* (London: Charles Knight, 1834), p. x; M. le Marquis de Laplace: *Traité de mécanique céleste*, tome cinquième (Paris: Bachelier, 1825), p. 350; facsimile reprint by [Editions] Culture et Civilisation (Bruxelles, 1967).

12. Translated from the Latin. The name of the author of the

review is not known. For the occasion of this particular comment, see § 6 *infra*.

13. Quoted from one of the most recent works to discuss this problem, Eric G. Forbes's Introduction to his annotated translation of *The Euler-Mayer Correspondence (1751-1755): A New Perspective on Eighteenth-Century Advances in the Lunar Theory* (London: Macmillan, 1971), p. 10.

More information is given below (in § 6) concerning this particular difference between Newtonian theory and observation.

14. See Wintner's treatise (cited in note 9 *supra*), Roberto Marcolongo: *Il problema dei tre corpi da Newton (1686) ai nostri giorni* (Milano: Ulrico Hoepli, 1919), and the historical notes in Forest Ray Moulton: *An Introduction to Celestial Mechanics* (ed. 2, New York: The Macmillan Company, 1914 [15th printing, 1962]).

15. In the Scholium following Prop. 35, Book III; see the discussion of this Scholium in § 6 *infra*.

16. On this topic, see David W. Waters: *The Art of Navigation in England in Elizabethan and Early Stuart Times* (London: Hollis and Carter, 1958); E. G. R. Taylor: *The Haven-Finding Art: A History of Navigation from Odysseus to Captain Cook* (London: Hollis & Carter, 1956); Charles H. Cotter: *A History of Nautical Astronomy* (London, Sydney, Toronto: Hollis & Carter, 1968); R. T. Gould: *The Marine Chronometer* (London: J. D. Potter, 1923), reprinted in facsimile (London, 1960).

17. These examples are taken from Saul Moskowitz: 'The Method of Lunar Distances and Technological Advance', *Navigation: Journal of the Institute of Navigation*, 1970, vol. 17, pp. 101–121; see also E. G. R. Taylor (*op. cit.*, n. 16 *supra*), p. 253.

18. These are described in the various works listed in note 16 *supra*.

19. The best detailed account I know of the principles and practice of the finding of longitude by the method of lunar distances is given in Charles H. Cotter: *A History of Nautical Astronomy* (cited in note 16 *supra*), pp. 205–242.

See, further, Eric Forbes's introduction to his edition of the Euler-Mayer correspondence (cited in n. 13 *supra*), pp. 16–21; and H.M. Nautical Almanac Office's publication 'A Modern View of Lunar Distances', *Journal of the Institute of Navigation*, London, 1966, vol. 19, pp. 131–153.

20. A major job of scholarly research needs to be undertaken on the methods actually used by table-makers in the eighteenth century, and on the accuracy of position those tables could give a user. For example, the *Theory of the Moon's Motion* (1702) appeared with the boast that the method would yield results good to 2 or 3 minutes, but would it?

Flamsteed, for one, didn't think so. Who was responsible for this value (2 or 3 minutes)? Was it Newton himself, or Halley or Gregory? This question occurs again below in § 6; but here it may be pointed out that these numbers do not occur in Newton's own text in the *Theory of the Moon's Motion* (1702), nor in the portion of that text (in the second and third editions of the *Principia*) incorporated into the Scholium following Prop. 35, Bk III. On this topic see note 43 to § 6 *infra*.

21. *Correspondence*, vol. 2, p. 481.

22. See the Koyré-Cohen-Whitman edition of the *Principia* with variant readings, cited in n. 12 to § 3 *supra*. The Scholium occurs in Book III, following Prop. 35; since this Scholium was replaced by the lengthy new one (in ed. 2, 1713), embodying some of the rules given in the 'Theory of the Moon', we may see how Newton had actually advanced on the subject from 1687 to 1700 (1702) and 1702 to 1713. And we may especially see the significance of the comments of the reviewer of the second edition, in the *Acta Eruditorum*; see n. 12 *supra*.

23. This statement was written in 1702, long before any public announcement had been made of the invention of the sextant (or reflecting octant). It certainly shows the author was aware that ordinary telescopes are not of much use for making observations from a ship at sea. Does this argue for Halley's authorship? Or is this point too elementary or too obvious to have necessarily required as author someone with Halley's experience at sea?

24. But readers of the *Theory of the Moon's Motion* (1702) would be told specifically (in the introduction) that Newton's results had demonstrated that the 'Irregularity of the Moon's Motion depends . . . on the Attraction of the Sun' (pp. iv–v); and they would most likely assume that the 'Rule' and 'Theory' by which 'the Place of the Planet [i.e., secondary planet, or moon] shall be truly Equated' (p. v) came directly from the theory of gravity. Those who encountered Newton's 'Theory of the Moon' in Gregory's textbook, in Harris's *Lexicon Technicum*, or in the *Miscellanea Curiosa* would not have been given such information.

6. The *Theory of the Moon's Motion* (1702) and the *Principia*

In the *Principia*, the theory of the moon's motion is introduced in two major different ways (first in Book I and then in Book III); Newton never consolidated his research on this topic into a single well-organized essay. The general principles of perturbation are presented initially in the famous Sec. 11 of Book I (so highly praised by Airy and others[1]), notably in Prop. 66 and its 22 corollaries. As mentioned earlier, the readers of the second and third editions would guess easily from the choice of letters that these corollaries were conceived in terms of a secondary planet (P) moving about the earth (T, for *Terra*) and being perturbed by the sun (S); but nowhere in this proposition and its corollaries, nor in any other part of Sec. 11, does Newton specifically refer to the moon. In Sec. 9 of Book I, 'Of the motion of bodies in moveable orbits; and of the motion of the apsides', the moon is mentioned but once: in a sentence which is not to be found in ed. 1, a fact that did not escape the sharp eyes of Bailly.[2]

This Prop. 66 of Book I is the closest Newton ever comes to the general problem of 'three bodies' which 'attract one another' with forces that 'decrease as the squares of the distances'; but Newton's consideration of the problem is subject to the restriction that there are 'two least bodies' that 'revolve about the greatest'.[3]

Newton's solution applies to motion in latitude (Cor. 10) and in longitude, and in particular to 'the annual equation, the motion of the apse line, the motion of the nodes, the evection, the change of inclination of the lunar orbit, and the precession of the equinoxes.'[4] Newton's results apply also to the phenomena of the tides. These purely theoretical results[5] are applied to the 'System of the World' in Book III, where Props. 22, 25–35 deal with the motion of the moon.[6] This presentation was greatly altered and expanded from edition to edition. A notable change appears in ed. 3 (1726), where two propositions occur following Prop. 33, on the mean motion of the sun from the nodes of the moon's orbit, and on the way to find the true motion of the moon's nodes from the mean motion, which were composed by John Machin. Among the new features of ed. 2 (1713) was the considerable amplification of Prop. 29 on the 'variation'. But, in the present context, the most significant and striking alteration made by Newton, in that part of Book III dealing with the moon's motion, was the completely new Scholium following Prop. 35, which appears for the first time in ed. 2. There was no other alteration or substitution of comparable magnitude in the material on the moon's motion from ed. 1 to ed. 2, or from ed. 2 to ed. 3.[7] And this new Scholium may particularly attract our attention here, since in it Newton gives a brief account of his own lunar research and summarizes or restates some of the new rules or 'equations' he had first published in the Latin textbook of David Gregory's, and in English in the same year in the *Theory of the Moon's Motion* (1702).

In the new Scholium, Newton stated explicitly that, 'By these computations of the lunar motions, I was desirous of showing that by the theory of gravity the motions of the moon could be calculated from their physical causes', a point he made equally clearly in a documentary history recently printed from Newton's MS.[8] In this essay, *Theoria Lunæ*, Newton presents some topics in a manner resembling the version given in the Scholium (in ed. 2); furthermore, some portions of the Scholium are taken over almost word for word, just as some paragraphs are almost identically the same as the Latin text of the 'Theory of the Moon' published by Gregory in 1702. With such a stress on gravity in these two documents (the essay and the Scholium), it seems more than a little odd that in the 'Theory of the Moon', Newton did not tell the reader that the rules contained therein had been derived in a new manner from a physical theory.[9] I use the expression 'new manner', since it was only with the publication of the *Principia* that it became possible, by application of the law of universal gravitation, to have a 'theory' of the moon's motion in the sense of a *'physical* theory' rather than what—according to Francis Baily—'may [today] be more properly called empirical or *tabular* theory'.[10] Baily actually went so far with this distinction that he designated 'the document, published in Gregory's work, by the title of Gregory's *Newtonian rules* for constructing lunar tables'; and he restricted 'the higher and more acknowledged title of *Newton's lunar theory* (that is, of his *physical* theory, in contradistinction to the empirical or *tabular* theory above mentioned) to the more modern exposition of the moon's motions deduced solely from his principle of Universal Gravitation.'[11]

Baily, arguing as the advocate for Flamsteed,[12] held that in the *Theory of the Moon's Motion* (1702), 'the alterations and amendments made by Newton [as far as regards the four principal inequalities of the moon[13]] were neither new nor (comparatively) very important'[14]; and so we may take rather seriously Baily's admission that 'Newton did not confine himself in that document, solely to the . . . ameliorations of the old tables; for he introduced 4 *new* equations to be added thereto'.

Following Baily's lead, let us concentrate our attention on the four new equations (or inequalities) presented by Newton in the 'Theory of the Moon', as published in 1702, and then see how they fared in 1713, in the second edition of the *Principia*. The subject of Newton's treatment of the four classical inequalities, together with an analysis of the whole corpus of his research on lunar theory, is at present a major area in Newtonian scholarship where research is needed—as has been mentioned more than once in this Introduction. Newton's new equations, as published in his *Theory of the Moon's Motion* (1702), include:

> an '*Equation of the Moon's mean Motion* depending on the Situation of her Apogee in respect of the Sun' (see p. 15; and Whiston's *Astronomical Lectures*, reprinted in facsimile below, p. 352); its maximum value occurs 'when the Moon's Apogee is in an Octant with the Sun'; this 'Equation,

when greatest, and Sun [is] *in Perigæo*', is 3'. 56", but 'if the Sun be *in Apogæo*, it will never be above 3'. 34".'

- an '*Equation of the Moon's Motion*, which depends on the Aspect of the Nodes of the Moon's Orbit with the Sun' (see p. 16; and Whiston, p. 354); its maximum value, is but 47", which (according to Whiston) 'is very small, but yet not altogether to be neglected.'
- a 'sixth Equation of the Moon's Place' depending on 'the Sine of the Sum of the Distances of the Moon from the Sun, and of her Apogee from the Sun's Apogee'; (p. 22; Whiston, p. 360); its maximum value is 2' 10'.
- a 'seventh Equation' depending on the 'Moon's [angular] Distance from the Sun' (p. 23; Whiston, p. 361); this equation, however, 'is encreased and diminished according to the Position of the Lunar Apogee' with respect to 'the Sun's'; and it is determined by a ratio with an apparent greatest value of 2' 20" which is not a true maximum but one which varies and so is actually a mean quantity, which can, in certain circumstances, be 'about 54". greater' or 'about as much less' and so 'librates between its greatest Quantity 3'. 14". and its least 1'. 26".'

With respect to the penultimate 'equation' in the above list, Whiston remarks (p. 361 of the facsimile reprinted below) that 'I must acknowledge my self to be altogether Ignorant' of how 'it should come to pass that this 6th Equation of the Moon should arise from Causes which are so unlike join'd together amongst themselves, as are the Motion of the Moon from the Sun and the Motion of the Apogee of the Moon from the Apogee of the Sun'. There was no 'Opportunity', he observes, 'for enquiring here in these Matters meerly Astronomical.' And then he says:

> In the mean while, I suspect that this Equation was rather deduc'd from Mr. *Flamsteed's* Observations, than

from Sir *Isaac Newton*'s own Argumentation; otherwise we had had two Equations assign'd each to its proper Cause out of this conjunct One. However, the Words of our Author are too clear to need that we should add one Word for interpreting them.

Here, once again, is a topic for research.

When Newton brought out the second edition of the *Principia* in 1713, incorporating these rules into the new Scholium to Prop. 35 of Book III, he introduced two most significant changes. Flamsteed was quick to note them both. In a letter to Abraham Sharp, written on 31 October 1713, Flamsteed declared 'that Sir Isaac Newton's sixth equation is not allowed by the heavens.' He then went on, 'He has lately published his *Principia* anew, wherein he makes this equation ablative where it was formerly to be added, and to be added where it was subductive; and has altered his seventh, so as in part to destroy it.' Flamsteed has 'not yet examined how this will answer'.[15]

In the Scholium, Newton explains that it was by using the theory of gravity that he had 'found that the annual equation of the mean motion of the moon arises from the varying dilation which the orbit of the moon suffers from the action of the Sun. . . .' And indeed, in cor. 6 to Prop. 66 of Book I, to which Newton refers us, the proof is given. He also 'found'[16] that 'the apogee and nodes of the Moon move faster in the perihelion of the Earth, where the force of the Sun's action is greater, than in the aphelion thereof, and that in the reciprocal triplicate proportion [i.e., inverse-cube ratio] of the Earth's distance from the Sun. And hence arise annual equations of those motions proportional to the equation of the Sun's centre.'

The third paragraph of the Scholium is essentially the same as the paragraph in the *Theory of the Moon's Motion* (1702), p. 15, describing (and giving the magnitude of) the first of Newton's new equations as in the foregoing list, without any significant difference, but with a wholly new introductory statement of some importance, reading:

> By the theory of gravity I likewise found that the action of the Sun upon the Moon is something greater when the transverse diameter of the Moon's orbit passeth through the Sun, than when the same is perpendicular upon the line which joins the Earth and the Sun: And therefore the Moon's orbit is something larger in the former than in the latter case. And hence arises another equation of the Moon's mean motion, depending upon the situation of the Moon's apogee in respect of the Sun. . . .[17]

Here, once again, Newton states expressly that the new equation arises from a condition disclosed from the 'theory of gravity'. But Newton does not make it explicit as to whether his conclusion concerning the increase and decrease of its quantity 'inversely as the cube of the Sun's distance' resulted from purely theoretical considerations, and—if so—what they were. This is yet another topic for Newtonian scholars to study.

The next paragraph in the Scholium is addressed to the second of the new equations in our list, and once again contains almost the identical content as the corresponding paragraph (p. 16) of the *Theory of the Moon's Motion* (1702). But here, too, Newton now has added an introduction stating expressly (but again without any details) that this equation took its rise from considerations of gravity. In the Scholium, he says by way of introduction:

> By the same theory of gravity, the action of the Sun upon the Moon is something greater, when the line of the Moon's nodes passes through the Sun, than when it is at right angles with the line which joins the Sun and the Earth. And hence arises another equation of the Moon's mean motion, which I shall call the second semi-annual. . . .

And whereas he had previously stated merely that the equation will 'when it is greatest amount to 47‴', he now declares:

> . . . and in the octants, where it is of the greatest magnitude, it arises to 47″. in the mean distance of the Sun from the

INTRODUCTION 61

Earth, as I find from the theory of gravity. In other distances of the Sun this equation, greatest in the octants of the nodes, is reciprocally as the cube of the Sun's distance from the Earth, and therefore in the Sun's perigee it comes to about 49″, and in its apogee to about 45″.

Once again, research is suggested as to the true role of gravity considerations in the identification and exact determination of this equation.

The succeeding three paragraphs of the Scholium differ in form and content from the corresponding presentation in the *Theory of the Moon's Motion* (1702). The third of these paragraphs, however, does conclude with a statement that replaces the earlier one of 1702, in two features, which may be seen at once by comparing the two printed below, side by side:

[*Theory of the Moon's Motion* (1702),] p. 22	[Scholium to Prop. 35, Book III, Motte's translation]
Again, as Radius to the Sine of the Sum of the Distances of the Moon from the Sun, and of her Apogee from the Sun's Apogee (or the Sine of the Excess of that Sum above 360°.):: so is 2′. 10″. to a sixth Equation of the Moon's Place, which must be *subtracted*, if the aforesaid Sum or Excess be less than a Semicircle, but *added*, if it be greater.	And as the radius to the sine of the angle thus found, ['by adding the distance of the Moon from the Sun, to the distance of the Moon's apogee from the apogee of the Sun'] so is 2′. 25″. to the second equation of the centre; to be added, if the forementioned sum be less than a semicircle, to be subducted if greater.

One difference is the numerical value used in the ratio, which has been increased from 2′ 10″ to 2′ 25″; but of far greater significance, as Flamsteed had been quick to observe, is the change in the conditions for addition and subtraction. The *Theory of the Moon's Motion* (1702) closely follows the text of Gregory's Latin book on astronomy of the same year, and Whiston (in 1707, in his *Prælectiones Astronomiæ*) merely reprinted the Latin text previously published by Gregory. The statement concerning the conditions for addition and for subtraction published in these three works is thus identical, and agrees furthermore with Newton's own MS version and Gregory's MS transcript of it.[18] But after 1713, when Newton had published the above-mentioned correction in ed. 2 of the *Principia*, there was no longer any excuse for continuing to reprint Newton's essay without alteration, as was done in both English editions of Gregory's textbook (and the second Latin edition), and the two English editions of Whiston's *Astronomical Lectures*, even though all declare in their second editions that the text has been 'corrected'. Nor was this correction introduced into the later reprintings of the *Miscellanea Curiosa* or of Harris's *Lexicon Technicum*; and it is not even mentioned as an annotation in Horsley's version in his edition of Newton's *Opera*.[19]

Whoever makes a careful comparison of Newton's *Theory of the Moon's Motion* (1702) and the revised Scholium will discern the correctness of Flamsteed's remark, quoted earlier, that Newton has so 'altered his seventh [equation] . . . as in part to destroy it': this equation is, to all intents and purposes, no longer a part of Newton's system. This example re-emphasizes the importance of making a detailed critical and analytical study of each successive stage of Newton's development of his own equations as well as his derivation of the older (or classical) inequalities from physical principles. And, in particular, only research among Newton's MSS will enable us to learn where or how the error arose in the sixth equation and how Newton came upon the correction, and also why and how the seventh equation became evanescent by the time of the Scholium. An incidental, but major, aspect of this research should be to illuminate the question raised by Whiston, as to Newton's actual use of gravity theory versus his direct dependence on observations and the fitting of geometric schemes to them, or even the charge made by Baily of Newton's reliance on Flamsteed's Horroxian model.[20]

In this regard, the *Theoria Lunæ*, long buried in Newton's papers, is of special interest. In introducing the geometric construction given in the Scholium following Prop. 35, Book III, ed. 2 of the *Principia*, Newton declares:

> Our countryman *Horrox* was the first who advanced the theory of the Moon's moving in an ellipse about the Earth placed in its lower focus. Dr. *Halley* improved the notion, by putting the centre of the ellipse in an epicycle whose centre is uniformly revolved about the Earth. And from the motion in this epicycle the mentioned inequalities in the progress and regress of the apogee, and in the quantity of eccentricity do arise.

In his presentation of a somewhat similar construction in the *Theory of the Moon's Motion* (1702), we find no such reference[21] made by Newton either to Horrox or to Halley.[22]

In the *Theoria Lunæ*, furthermore, Newton put together a brief history of some prior researches as an introduction to an account of his own achievement. He begins by referring to the ancient notion of 'excentric circles in solid orbs', and to Tycho's cometary studies as the source of the discovery 'that the heavens are not solid' (somewhat similar to a beginning statement in the first version of his 'System of the World').[23] Newton then says that although Kepler had discovered that planets move in elliptical orbits around the sun, it was Horrox who introduced both the elliptical orbit for the motion of the moon and the law of areas for motion in that elliptical orbit.[24] Newton was wrong on both counts. Kepler himself had written about an elliptical lunar orbit, as Edward Rosen has shown,[25] and Horrox never discussed the law of areas in any connection —neither for the motion of the planets nor for the motion of the moon.[26]

In the *Theoria Lunæ*, Newton gave an accurate description of Horrox's chief contribution to the lunar theory, a motion along an ellipse whose 'excentricity does not remain the same as in the primary Planets, but increases and diminishes in turn, in accordance with the position of the Moon's Apogee with respect to the Sun.'[27] Then, in this essay on the *Theoria Lunæ*, as in the new Scholium, Newton refers to Halley's innovation:

> ... And Halley from astronomical observations gathered that the superior focus of the Moon's Orbit is carried around with uniform motion on the circumference of a circle whose radius is the semidifference of the greatest and least excentricity and whose centre revolves with uniform motion about the Earth at a distance equal to the mean excentricity.[28]

This particular innovation of Halley's has all but dropped out of the literature of the history of astronomy, and is not even mentioned in the two recent biographies of Halley.[29]

Halley, in fact, made two contributions to the lunar theory. The first, the one mentioned by Newton, was published as a tract appended to his catalogue of stars in the Southern Hemisphere, *Catalogus Stellarum Australium* (London, 1679).[30] In referring to it in his textbook on astronomy, David Gregory said, 'This aforesaid elliptic Figure of the Orbit of a Satellite was observ'd in the Moon by that most ingenious Astronomer Dr. *Halley*, who in the year 1679, first of any Body, publish'd it in his Treatise concerning the Emendation of the Moon's Theory.'[31] Halley's second discovery was the secular acceleration of the moon's motion.[32]

Toward the end of the first paragraph of the MS essay, *Theoria Lunæ*, Newton begins to discuss his own work with these words, 'Nos in præcedentibus ...', referring to his applications of the theory of gravity ('Theoriam gravitatis') to celestial phenomena, both the investigation of the causes of the motions and the computation of the quantities of the motions from those causes. Hence it would appear that the essay would have been intended to follow the first 35 propositions of Book III of the *Principia*, and probably is a prior draft of the new Scholium that appears following Prop. 35 in the second edition. Accordingly, this would place the text in time between the 'Theory of the Moon' and the new Scholium, no doubt much nearer the latter. This view is strengthened by the fact that the *Theoria Lunæ* contains much the same material concerning Horrox and Halley as does the Scholium (though more detailed), and it has two paragraphs (penultimate and ante-penultimate) which are all but identical to two paragraphs in the Scholium.

In addition to the historical information and details concerning his own achievement, Newton's *Theoria Lunæ* is noteworthy for two statements. One occurs in the final paragraph, where Newton introduces a topic not explicitly discussed in the Scholium at all.[33] The other is found at the beginning of the second paragraph, in a reference to Flamsteed, whose name is conspicuously absent from this Scholium.[34] Newton is saying that it was 'by deducing the celestial motions from the laws of gravity' that he had also found that 'the annual Equation of the mean motion of the Moon . . . arises from the various dilations of the orbit of the Moon by the force of the Sun'; and then he says that both Kepler and Horrox had 'composed' this annual equation with the equation of time, whereas Flamsteed had treated it separately.[35]

A truly satisfying view of Newton's lunar theory and its significance must be built upon three foundations, not one of which is at present in a state worthy of the subject. The first of these is a sound view of the state of lunar theory from Tycho Brahe[36] to Flamsteed, including the calculation of tables of the moon's motion and position and their uses. In particular, it would be important to have thorough and reliable critical analyses of the lunar researches of Kepler, Horrox, and Flamsteed, especially of the work on the moon by such relatively minor figures as Bullialdus, Wing, and Streete: not only because they wrote about the moon's motion, or published tables, but because they were read by Newton during his formative years. A second major set of tasks is to trace the development of the lunar theory from Newton to Laplace, including not only the major figures (Euler, Clairaut, Tobias Mayer,[37] d'Alembert, Lagrange) but also the table-makers of the post-Newtonian era—especially to find out whether or not, or to what degree, these table-makers actually made use of Newton's rules, published in the *Theory of the Moon's Motion* (1702) or later reprints in English, or in the Latin printings.[38] Finally, a thorough and accurate critical history of Newton's own researches is the third and necessary foundation for our understanding of the lunar theory.[39]

Newton's views concerning the motion of the moon have long been available in print in the three editions of the *Principia*; plus the first version of what was to become Book III of the *Principia*, published posthumously in both Latin and English versions and misleadingly given the name of the subtitle of Book III, *Systema Mundi* or 'System of the World'. These may be supplemented by his *Theory of the Moon's Motion* (1702), reprinted below in its first Latin printing and in two English versions. Additionally, Newton's notes in relation to possible revisions of the *Principia* exist in MS in the University Library, Cambridge, Portsmouth Collection, MS Add 3965. A large collection of MS notes and calculations, together with MS lunar tables, MS Add. 3966, is of the greatest importance;[40] the *Theoria Lunæ*, now in print, comes from this latter group of MSS. Yet additional information is available in MS and partly in print in Newton's correspondence—notably with Halley, Flamsteed, and Cotes. The volume of Newton-Cotes correspondence in Trinity College Library (MS R.16.38) needs to be studied, along with Edleston's notes to his edition of these exchanges.[41] Two additional jobs needing to be done are to make a thorough search of the Gregory MSS, chiefly in the Edinburgh University Library, for possible documents that might illuminate Newton's lunar theory; Gregory's comments on the portions of Newton's *Principia* relating to the moon in his MS *Notæ in Newtoni Principia Mathematica Philosophiæ Naturalis* are also worthy of study. Finally, Newton's early notebooks, and the books he read, should be examined for sources of his knowledge of the moon.[42]

In view of the importance of the theory of the moon's

motion as a principal fruit of the theory of gravitation and a witness to the power of the Newtonian celestial mechanics,[43] to say nothing of the practical importance of this subject for navigation, our present unsatisfactory state of knowledge must be taken as an index of the current inadequacy of studies in the history of the exact sciences.[44]

What would be most desirable would be to use the aforementioned printed and manuscript sources to produce a coherent presentation of Newton's theory of the moon—historical and analytical and critical—indicating both its real achievements and its failures. It would be of the greatest importance to scholars to learn just how (or to what extent) the separate and individual propositions, of Book III of the *Principia* dealing with one or another aspect of the moon's motion, are either related to one another or are in fact dependent on the general dynamical principles explored in Book I. In particular, it must be said again and again, we just don't know for certain how much of the theory of the moon presented by Newton in Book III of the *Principia* and in the pamphlet on the *Theory of the Moon's Motion* (1702), truly came from his gravitational theory, and how much was in the tradition of epicyclic geometric constructions, tailored to fit the needs of new and successively more complete and more accurate data concerning lunar positions and lunar motions. It must be confessed that even the parts of Newton's propositions on the moon's motion that seem to be most intimately connected with his gravitational theory need to be examined critically in order to reveal the extent of the 'analytical' development of the subject that may be masked by a geometric form.[45] In short, before we can properly and truly evaluate a Newtonian theory of the moon's motion and its influence, we must know exactly what it is.[46]

Notes
1. See § 5 *supra*, esp. note 5.
2. It was Newton's plan, in Book I of the *Principia*, to introduce 'here and there', as he said, 'some philosophical scholiums' to illustrate

the general or mathematical principles (which of and by themselves would otherwise have seemed 'dry and barren'); it is thus particularly noteworthy that in Sec. 11 of Book I he did not introduce a single scholium suggesting a physical situation to which the propositions or their corollaries might apply.

At the end of the penultimate paragraph of Prop. 45 (the third and last proposition of Book I, Sec. 9), Newton computes that in the orbit he is studying, 'the upper apsis in each revolution will go forward 1 deg. 31 m. 28 sec.'. Then, he observes 'The apsis of the Moon is about twice as swift.' This is the way the *Principia* reads in ed. 3 (1726).

In ed. 1 (and in the MS from which ed. 1 was printed) the numerical value was given as 1 degree, 31 minutes, 14 seconds. In the first two editions, the discussion concluded at that point. It was not until ed. 3 that the comment ('Apsis lunæ est duplo velocior circiter') was added; but in both copies of ed. 2 in which Newton entered emendations for a future edition, this additional sentence was given as 'Quæritur an hic motus a dupla vi extranea oriri possit.' See further, the edition of the *Principia* with variant readings (cited in note 12 to § 3 *supra*), vol. 1, p. 242, line 30.

The comment by Jean-Sylvain Bailly is given in vol. 2 of his *Histoire de L'astronomie moderne* (Paris, 1785), p. 508, note (a): 'La prop. XLV du premier livre du *Principes* donne le mouvement des absides, qui résulte d'une addition à la force centrale: Newton n'en a point fait d'application à la lune, du moins dans la première édition; dans les suivantes il a montré une supposition de force, qui donneroit la moitié du mouvement de l'apogée (*Edit. Franc.* T. I, p. 151). . . .'

Rouse Ball (*Essay*, cited in n. 10 to § 5 *supra*), p. 85, observed: 'The second corollary [to Prop. 45, Book I] is illustrated by showing that if this extraneous addition by $\frac{1}{357.45}$th part of the force under which the body would revolve in the ellipse, then in each revolution the apse line would progrede $1°\ 31'\ 14''$, a number which in the second edition was corrected to $1°\ 31'\ 28''$. It seems clear from the Portsmouth papers that this was given merely as an illustration of the method, but in the third edition the words 'Apsis lunæ est duplo velocior circiter' were added. It may be that this remark was inserted in order to show that the corollary was not applicable to the case of the moon—in fact only one part of the sun's disturbing force is here treated—but a reader might also think that the remark was intended to point out a discrepancy between the theory and observations. As Newton had explained the similar difficulty in the case of the node, some writers suspected (*ex. gr.* Godfray, in his *Lunar Theory*, second edition, 1859, art. 68) that the scholium in the first edition to book iii. prop. 35 meant that he had found the explanation: but nowhere in the *Principia* does Newton explicitly give this explanation, though in book iii. prop. 25 he estimates that the total disturbing force of the sun on the moon bears to the earth's centripetal force the ratio 1 to $178\frac{29}{40}$, which would make the annual progression of the apse line about what it actually is. The remark at the end of book i. prop. 45 was, however, read by many as indicative of a variance between observation and the Newtonian theory, and the explanation of a difference which had become an obstacle to the universal acceptance of the Newtonian system was first given by Clairaut. The Portsmouth papers contain Newton's original work, and show that he had found, by carrying the approximation to a sufficiently high order, that the mean annual motion of the apse line was $38° \ 51' \ 51''$, which is within $2°$ of the true value. . . .' Here is a major topic for further research, especially since Rouse Ball's opinion concerning the words 'Apsis lunæ est duplo velociter circiter' seems wide of the mark. See note 46 *infra*.

3. An admirable analysis of Prop. 66 and its 22 corollaries is given in Henry Lord Brougham and E. J. Routh: *Analytical View of Sir Isaac Newton's Principia* (London, 1855; facsimile edition, New York, London: Johnson Reprint Corporation, 1972), pp. 126–134; based upon a review of Laplacian principles, pp. 108 sqq. Brougham declares (p. 134): 'Sir Isaac Newton here deduces most of the leading disturbances in the motions of three bodies, for example, the moon, earth, and sun. . . . We perceive in succession the motion of longitude and latitude; the various annual equations, motion of the apsides (in which, however, by omitting the consideration of the tangential force, he calculated the amount at one half its true value), the evection, the alteration, and inclination; the motion of the nodes.'

Observing that the 'greater part of the Third Book is occupied with the application of these corollaries to the actual case of the moon, earth, and sun', Brougham concludes that 'it is not any exaggeration to affirm that the great investigations which have been undertaken since the time of Sir Isaac Newton . . . are an application of the improved calculus to continue the inquiries which he thus here began.'

(I have attributed both statements to Lord Brougham, since the first half of the book is an unaltered reprint of an earlier essay of his; see p. x of the editor's introduction to the reprint edition mentioned above.)

4. Quoted from the outline of the *Principia* given by W. W. Rouse Ball in ch. vi of his *Essay on Newton's 'Principia'* (cited in n. 10 to § 5 *supra*).

5. As Newton explained in the introduction to Book III of the *Principia*, his plan was to develop the mathematical principles in Book I (and in Book II), with a scholium here and there to suggest possible applications to physical situations, as has been explained above, in note 2. Then, in Book III, these principles are drawn upon, to yield information about the motion of planets and their satellites, our moon, and comets; and also to explain the production of tides.

6. Further propositions about the moon occur in Book III. Prop. 37 has as its subject the magnitude of the lunar tide, from which Newton gets the mass and density of the moon, in a result wide of the mark. In ed. 2, the discussion of Prop. 37 was considerably expanded, and three corollaries were appended; while in ed. 3 two further corollaries were added. Prop. 38 discusses the spheroidal shape of the moon, and the reason why the moon always shows her same face to the earth. The succeeding 3 lemmas and 'Hypothesis 2' (formerly 'Lemma 4') lead to Prop. 39 on the moon's action on the earth's equatorial bulge to produce precession. Prop. 23 deals with the ways in which the inequalities in the moon's motion lead to a knowledge of the motions of the moons of Jupiter and of Saturn, while the subject of Prop. 24 is the production of the tides by solar and lunar forces.

7. Even the new propositions by Machin in ed. 3 yield results, as Rouse Ball observed, that are 'consistent' with Newton's, even though the 'methods differ from those of Newton'.

8. This essay, taken from Newton's autograph version in the University Library, Cambridge, MS Add. 3966, is printed in vol. 4 of Newton's *Correspondence* (1967), pp. 1–8; it is entitled *Theoria Lunæ*.

9. On this point, see n. 24 to § 5 *supra*.

10. Baily, *op. cit.* (note 3 to § 1 *supra*), p. 690. This distinction between the two usages (of the word 'theory') has already been introduced in §1 *supra*.

11. Baily, *ibid.*, p. 691.

12. I do not care to enter into the century-old debate as to whether Baily's efforts to restore justice to Flamsteed made him so partisan as to render his judgments concerning Newton of little value, if not downright wrong.

13. These are the equation of the centre (apparently found by Hipparchus), the evection (discovered by Ptolemy), the variation (discovered by Tycho Brahe), and the annual equation (also due to Tycho).

On this topic see Victor Eugene Thoren: 'Tycho Brahe on the Lunar Theory' (Ph.D. Thesis, Indiana University: June, 1965). Thoren has published four articles based upon his thesis, as follows:

'Tycho Brahe's discovery of the variation', *Centaurus*, 1967, vol. 12, pp. 151–166; 'An early instance of deductive discovery: Tycho Brahe's lunar theory', *Isis*, 1967, vol. 58, pp. 19–36; 'An "unpublished" version of Tycho Brahe's lunar theory', *Centaurus*, 1972, vol. 16, pp. 203–230; 'Tycho and Kepler on the lunar theory', *Publications of the Astronomical Society of the Pacific*, 1967, vol. 79, pp. 482–489.

14. The equation of the centre arises from the fact that the earth is not at the centre of the moon's orbit; this inequality is also known as the 'elliptic inequality'. The apse, however, is not stationary, but moves with a speed of about $3°$ per lunar revolution of $360°$ (as Hipparchus found and 'all modern observations confirm'). Hugh Godfray, whose book on the lunar theory remains today one of the best introductions to Newton's work (translated into the analytic language of differential equations), commented as follows on the difference between Newton's computed value of about $1\frac{3}{5}°$ rather than about $3°$ per revolution:

'Newton himself was aware of this apparent discrepancy between his theory and observation; and we are led, by his own expressions (Scholium to Prop. 35, lib. III. in the first edition of the *Principia*), to conclude that he had got over the difficulty. This is rendered highly probable when we consider that he had solved a somewhat similar problem in the case of the node; but he has nowhere given a statement of his method: and Clairaut to whom we are indebted for the solution, was on the point of publishing a new hypothesis of the laws of attraction, in order to account for it, when it occurred to him to carry the approximations to the third order, and there he found the next term ... [to be] nearly as considerable as the one already obtained. ... [This work resulted in] reconciling theory and observation, and removing what had proved a great stumbling-block in the way of all astronomers.' See pp. 67–68 of Godfray's book, cited in n. 5, §5 *supra*. For a commentary on Godfray's assumption that Newton's Scholium to Prop. 35, Book III, in the first edition 'meant that he had found the explanation', see Rouse Ball's *Essay* (cited in note 10 to § 5 *supra*), p. 85; and the extract in n. 2 *supra*.

Godfray correctly observes (p. 68) that the 'motion of the apse line is considered by Newton in his *Principia*, lib. I., Prop. 66, Cor. 7'; that 'Newton has considered the evection, so far as it arises from the central disturbing force, in Prop. 66, Cor. 9, of the *Principia*'; that 'As far as terms of the second order, the coefficient of the variation is independent of e the eccentricity, and k the inclination of the orbit. It would therefore be the same in an orbit originally circular, whose plane coincided with the plane of the ecliptic: it is thus that Newton has considered it. *Princip.* prop. 66, cor. 3, 4, and 5.' Furthermore, 'The annual equation is, to this order, independent of the eccentricity and inclination of the moon's orbit, and therefore, like the variation, would be the same in an orbit originally circular. *Vide* Newton, *Principia*, prop. 66, cor. 6.' For Godfrey's further comments on the 'longitude of the node' and evection in latitude (in relation to '*Principia*, book III, props. 33 and 55'), see his p. 77; see also p. 80 for the variation, where 'the moon's orbit would be an oval having its longest diameter in quadratures and least in syzygies. *Principia*, lib. I. prop. 66, cor. 4.'

Clearly another topic for research is to find out whether (or to what degree) Baily may or may not have been correct in his judgment concerning the worth of Newton's 'alterations and amendments' to the known inequalities—other than, of course, the possibility of deriving them mathematically from physical principles.

15. Before long, Flamsteed reported to his correspondent: 'I told you that the heavens rejected that equation of Sir I. Newton, which Gregory and Whiston called his sixth. I had then compared but 72 of my observations with the tables: now, I have examined above 100 more. I find them all firm in the [*rejection* of the] same, and in the seventh too. And whereas Sir I. Newton has in his new book (pages 424 and 425) thrown off his sixth, and introduced one of near the same bigness but always of a contrary denomination, and a bigger in the room of the seventh, if I reject them both, the numbers will agree something better with the heavens than if I retain them: so that I have determined to lay these *crotchets* of Sir I. Newton's wholly aside.' Quoted from Baily's *Flamsteed* (cited in note 3 to § 1 *supra*, pp. 309, 698).

16. In this paragraph, he says, 'Further I found . . . ' ('Inveni etiam quod . . .'); but he does not repeat the explicit reference to the theory of gravity, which is clearly implied.

17. This, and the foregoing, translations from the *Principia* are based on Andrew Motte's version.

18. See § 3 *supra*.

19. See § 2 *supra*. It is a cause for wonder that Newton's rules for the moon's motion continued to be reprinted throughout the century without his corrections. The reason may be that the modifications made in lunar theory after Newton's death (Clairaut, Euler, etc.) were of such consequence as to make Newton's own work of interest only in a theoretical or historical sense. Here is another matter for research.

20. Horrox's lunar theory was sketched out in two letters to Crabtree, in a manner so sketchy and fragmentary that these letters were to be omitted from the edition of Horrox's *Opera Posthuma*. Flamsteed,

concerned with the moon's motion, and having found that 'Bullialdus's, Wing's, and Street's theories were erroneous', was delighted to find that he could construct or reconstruct a Horroxian theory from the notes and rules (in the letters to Crabtree) that was 'near the truth'. Horrox's scheme or theory considered 'the moon to move in an ellipse with a variable eccentricity, and having a libratory motion of the apsides' (Baily's *Flamsteed*, pp. 31, 680). For Flamsteed's improvements of the theory, and the bibliographical details concerning the insertion of Flamsteed's edited Horroxian lunar theory into the *Opera Posthuma* (of which some copies have an alternative title-page, *Opuscula Astronomica*), and the duplication of pages 465–470 in most copies, see Baily's *Flamsteed*, p. 681. On pp. 681–683, Baily gives a convenient summary of the Horroxian theory, with diagrams.

For further bibliographical information regarding Horrox's *Opera Posthuma*, see John E. Bailey, 'The writings of Jeremiah Horrox and William Crabtree', *The Palatine Note-Book*, 1883, vol. 3, pp. 17–22; an essentially identical article by Bailey with approximately the same title also appeared in *Notes and Queries*, 6th series, 2 December 1882, vol. 6, pp. 441–44. See also Rigaud, *Correspondence . . .*, vol. 2 pp. 149–50, note 5.

21. In point of fact, Newton mentions no names whatever in the series of rules printed in 1702; it should be kept in mind that, unlike a Scholium written for the *Principia*, or an essay like the *Theoria Lunæ* printed below, these rules are stated tersely and directly, without any comments or an introduction or conclusion by Newton. It must have been only the zeal of a partisan that caused Baily (*op. cit.* pp. 370, 735n.) to have been critical of Newton for not having referred to, Flamsteed in the text published by Gregory and translated in *Theory of the Moon's Motion* (1702). Baily might more justly have taken Newton to task for not having mentioned Flamsteed in the new Scholium to Prop. 35 of Book III.

22. Edleston (*op. cit.*, n. 3 to §1 *supra*), p. lxx, n. 143, would rebut the 'renewed assertion' made by Baily, in his *Supplement* (p. 735) that 'there is not a single allusion made to Flamsteed' by observing that in the English reprints (*Theory of the Moon's Motion* (1702) Harris's *Lexicon Technicum*, the *Miscellanea Curiosa*), 'the mention of Flamsteed's name comes *after* the title of the tract, not *before* it as in Gregory's *Astronomy*.' This is, alas, on a level not worthy of the learned and judicious notes that are found in Edleston's volume. And he was simply wrong in his final guess: 'It is extremely improbable that the essay was communicated to Gregory in the naked form in which it stands within inverted commas in his *Astronomy*: it must have been accompanied by some notice of Flamsteed's Observations and their

near agreement with the results derived from the Theory, the substance of which Gregory chose to embody in an introductory paragraph, then prefixing the title "Lunæ Theoria Newtoniana," and finally giving us the actual Theory in its author's own words—a bare numerical statement of facts and rules. . . .'

23. In that work, Newton had said that 'the phænomena of Comets can by no means consist with the notion of solid orbs'; but he did not refer to Tycho specifically, as he does (quite correctly) in this connection in the *Theoria Lunæ*. The 'System of the World' was the name given to an early version of Book III of the *Principia*, which was published posthumously. The contemporaneous English translation, *A Treatise of the System of the World* (London, 1728, 1731), has been reprinted in facsimile by Dawsons of Pall Mall (London, 1969).

24. In the new Scholium following Prop. 35, Book III (*Principia*, eds. 2, 3), Newton merely says, 'Our countryman *Horrox* was the first who advanced the theory of the Moon's moving in an ellipse about the Earth placed in its lower focus'; he does not then say, as in the MS *Theoria Lunæ*, that Horrox also declared that the moon moves in that orbit according to the law of areas.

25. See Edward Rosen: 'The Moon's Orbit in Kepler's *Somnium*', *Centaurus*, 1966, vol. 11, pp. 217–221.

Newton, however, apparently never read Kepler's *Somnium*. A brief outline of the development of Kepler's views concerning an elliptical lunar orbit is given by Victor E. Thoren in his (as yet) unpublished doctoral dissertation (see n. 13 *supra*), pp. 179–180.

26. At least, I have never found a direct reference to the area law in Horrox's *Opera Posthuma* (London, 1773), a copy of which was given to Newton by John Collins.

27. See note 20 *supra*.

28. In Propr. 29 of Book IV of Gregory's textbook (the proposition which is followed by a scholium containing Newton's 'Theory of the Moon'), pp. 559–560 of the English edition of 1726, there occurs 'a short Account of such of the foremention'd Corrections concerning the Moon's Motion as have hitherto been taken Notice of by the Constructors of Tables, and such of them as have been neglected.' First of all, 'they neglect all the Annual Æquations or Corrections depending upon the various Distances of the Earth from the Sun'; accordingly, they 'consider only that which arises from the encreas'd and diminish'd Periodical Time of the Moon about the Earth, from the diminish'd or encreas'd Distance of the Earth from the Sun; that is, which depends upon the Earth's Mean Anomaly'. Gregory criticized Tycho, Kepler, and Horrox for confounding 'this inequality'

(which they had all 'observ'd') with 'the others'. According to Gregory Tycho 'quite left out that Part of the Æquation of Time which depends upon the Earth's Mean Anomaly, upon which also this Inequality of the Moon depends.' Kepler and Horrox, however, 'having carried the Matter farther for their equating of Time, make use of an additional Physical Part, when the true and Astronomical one may be taken away, and so on the contrary.' But it was the 'most ingenious Dr. *Halley* [who] was the first that separated this Inequality (which Mr. *James Gregory* suspected to belong to the Moon . . .) from the others, in a Treatise subjoin'd to his *Catalogue of the Southern fixed Stars*.'

Baily (*Flamsteed*, p. 683; cf. pp. 134, 137) asserts that the epicycle construction 'adopted' by Newton for 'the variation of the eccentricity of the lunar orbit' comes from the one given by Flamsteed in his '*Epilogue* to the tables', with a 'slight correction of the eccentricity [which] was imparted to Newton as a secret by Halley'.

29. This contribution is not mentioned in either Colin A. Ronan: *Edmond Halley: Genius in Eclipse* (Garden City, New York: Doubleday & Company, 1969) or Angus Armitage: *Edmond Halley* (London: Nelson, 1966). Nor is there a reference to this topic in Eugene Fairfield MacPike (ed.): *Correspondence and Papers of Edmond Halley* (Oxford: at the Clarendon Press, 1932).

30. I owe this reference, initially, to Prof. A. R. Hall. Mr Craig Waff informs me that at the 'Observatoire' in Paris, he 'came across another edition of this work which had French and Latin texts running concurrently: Edmond Hallai [*sic*] *Catalogue des estoilles australes* . . . (Paris: chez Jean Baptiste Coignard, 1679)'.

Baily (*Flamsteed*, p. 684) notes in passing that by 'this new theory of the libratory motion of the apsides, and the variation of the eccentricity', Horrox 'proposed to unite the two principal equations of the lunar irregularities; namely, the Equation of the centre and the Evection'. Since 'the Annual equation was got rid of, by a supposed equivalent alteration in the Equation of time', there were needed but two 'tables for finding the corrections of the moon's place in her orbit'. Baily then observes: 'In a short tract published by Halley at the end of his *Catalogus Stellarum Australium*, 1679, entitled *Quædam lunaris theoriæ emendationem spectantia*, page 12, he ventures an opinion, that upon Horrox's theory it would be practicable to expound also the law of the Variation, by supposing the moon's orbit, in the line of its sizygies, to be compressed 1/90th part towards the earth, and thus reduce these equations to *one* only: but he says that he leaves it to others to follow out the plan.'

Baily points out, furthermore, that 'Flamsteed was aware of the inaccuracy of this substitution' (which eliminated a separate correction for the annual equation), but he retained it 'at first . . . out of compliment to Horrox'.

31. *Op. cit.* (see n. 6 to § 2 *supra*), vol. 2, sec. 1 of Book IV, p. 471; Gregory then observes, 'But the History of the Academy of Sciences in *France*, published in 1698, says, that Mr. *Picard* had made the same Observation long before, namely in 1668.'

Convenient summaries of Halley's lunar theory of 1679 are given in the following works: Jérôme de Lalande: *Astronomie*, tome second (Paris, 1764), p. 171, § 1451, (seconde édition, revue et augmentée, Paris, 1771), p. 219, §§ 1450–1451; Jean-Baptiste Joseph Delambre: *Histoire de l'astronomie au dix-huitième siècle* (Paris, 1827), pp. 118–119, 123. Information on this subject by A. Rupert Hall will appear in vol. 5 of Newton's *Correspondence*, of which he is now the chief editor.

32. Soon after the publication of the *Principia*, in 1693, Halley discovered a new lunar phenomenon, known as the secular acceleration of the moon's mean motion. By a careful comparison of ancient and modern eclipses, he found that the period of the moon's mean revolution had been decreasing since the days of ancient Mesopotamia.

33. This paragraph deals with the gravitational action of the sun on the moon, which is greater 'when the Moon's apogee and the Sun's perigee are in conjunction than when they are in opposition'—which results in 'two periodic equations: one of the Moon's mean motion, the other of the motion of its Apogee'. This paragraph may be compared to pp. 15 and 24 of the *Theory of the Moon's Motion* (1702).

34. One basic difference that kept Newton and Flamsteed apart was Newton's belief that a perfected lunar theory would have as great a value as testimony to the truth of the gravitational principles as it would for the prediction of the moon's motion or position (for its own sake or for use in navigation); whereas Flamsteed did not care overly whether the theory of gravitation were validated or not, and his primary concern was to improve observations and to find any scheme or set of rules that would closely agree with observation. On this topic see I. B. Cohen: *Introduction to Newton's 'Principia'* (cited in n. 10 to § 3 *supra*), pp. 172 sqq., and Baily's *Flamsteed, passim*. As is made clear in the correspondence with Flamsteed in the 1690's, Newton was aware that 'the theory of the moon' was 'very intricate', but he was certain that in the perfecting of the theory of the moon's motion, 'the theory of gravity [is] so necessary to it, that I am satisfied it will never be perfected but by somebody who understands the theory of gravity as well, or better then I do.'

For further information concerning the relations between Newton and Flamsteed in the 1690's, see the most recent edition of their correspondence, in vol. 4 of Newton's *Correspondence* (Cambridge, 1967). And see notes 21, 22 *supra*.

35. See n. 30 *supra*.

36. Tycho's contribution has been studied by Victor Thoren; see n. 13 *supra*.

37. See Eric G. Forbes, 'Tobias Mayer's contributions to the development of lunar theory', *Journal for the History of Astronomy*, 1970, vol. 1, pp. 144–145.

38. A bare beginning has been made by Baily (*Flamsteed*, pp. 701 sqq.) in finding out how widely Newton's 'Theory of the Moon' was read and actually used by table-makers, of whom he finds the first to have been Wright (1732) and Leadbetter (1735). But Baily did not look into the specific discussions of Newton's rules (published first in 1702) as distinct from his Scholium to Prop. 35, Book III, in the second and third editions of the *Principia*. Mr Craig Waff, of the History of Science Department of Johns Hopkins University, has been devoting himself to the history of the lunar theory in the post-Newtonian century, and has sent me a preliminary list of lunar tables 'which I believe incorporated Newton's theory in their construction' (personal communication, 5 Jan. 1971). Mr Waff now announces that his doctoral dissertation, in progress, is on the subject: 'Universal Gravitation and Lunar Theory 1687–1767'. A chapter will be devoted to printed and MS tables which used Newton's theory.

39. In Livre XVI of his *Mécanique céleste* (tome cinquième, Paris, 1825—facsimile reprint: Culture et Civilisation, Brussels, 1967), Laplace gives (Ch. 1) an admirable summary of the history of lunar theory, including a succinct but thorough account in words of the lunar doctrine expounded in the *Principia*. Laplace says (p. 352), *en passant*, 'Cette méthode de Newton est fort ingenieuse; et l'on verra dans le chapitre suivant, qu'en traduisant en analyse, elle conduit facilement aux équations différentielles du mouvement lunaire.' This judgment is well exemplified in the following ch. 2, 'Sur la théorie lunaire de Newton'.

40. Some aspects of Newton's proposed revision of the lunar theory were written up by J. C. Adams and published in the Appendix to the Preface to *A Catalogue of the Portsmouth Collection of Books and Papers written by or belonging to Sir Isaac Newton* (Cambridge: at the University Press, 1888), and are partially reprinted in Rouse Ball's *Essay* (cited in n. 10 to § 5 *supra*), p. 109, 126 sqq.

41. On 16 Feb. 1711/12 Cotes wrote to Newton about a discrepancy in the motion of the 'Earth's Aphelium' in the Scholium proposed to Prop. XIV of Book III and 'Your new Theory published by Dr Gregory' and wondered 'whether these numbers are propos'd barely

as an Example'; see Edleston's edition of the Newton-Cotes correspondence (cited in n. 3 to § 1 *supra*), p. 64, and cf. p. 66 for Newton's reply on 19 Feb. On 3 May 1712, Cotes discussed the new Scholium to Prop. 35; Edleston's commentary on this new Scholium (pp. 109 sqq., following the printing of Cotes's letter) is more than usually valuable. Unfortunately, however, Edleston did not print the first version of the Scholium *in extenso*, although he did outline some of the main points which Newton later revised in response to Cotes's criticisms. The documents are to be presented more fully in vol. 5 of Newton's *Correspondence* (now in progress). See also, in Edleston's volume, the letters and editorial comments on pp. 120–141. A comment by Cotes, 'I quæry whether *Halleius superiorem Ellipseos umbilicum in Epicyclo locavit* should not be also chang'd into *Halleius centrum Ellipseos*', is accompanied by the statement that 'I have not Dr Halley's little Treatise by me concerning the Lunar Theory' (Edleston p. 121; Cotes to Newton, 10 Aug. 1712). Cotes also observes that, 'The Æquation which You here call *Æquatio centri secunda* is I perceive the same with that which in Dr Gregories Astronomy You call *Æquatio loci Lunæ sexta*'; he would be 'very glad' to learn from Newton 'the reasoning by which it is established'. The remainder of the correspondence on this subject (pp. 122–141) confirms the impression that Cotes made a major contribution to the presentation of the lunar theory in the second edition of the *Principia*.

42. A detailed analysis of Newton's MSS may hopefully reveal Newton's methods, and in particular show the extent of his reliance on Horrox: thus to test the validity of Baily's severe judgment (*Flamsteed*, pp. 691–692) that at least a part of Newton's principles or rules 'ought in strictness to be called *Newton's Horroxan* theory; the term, in fact, which Flamsteed adopts in his letter to Dr. Wallis'.

43. Displayed prominently at the head of all printings of Newon's 'Theory of the Moon' is a statement concerning a degree of accuracy of $2'$ or $3'$; this boast is properly Newtonian, for it accords with a letter from Newton to Flamsteed (cf. Baily, *Flamsteed*, p. 687), and it was apparently circulated by Halley and Gregory. But in the *Principia* and in his own part of the essay printed by Gregory (as also in Newton's essay, 'Theoria Lunæ'), Newton was careful to avoid an explicit reference to the degree of accuracy of his rules. Thus a research problem exists: to find out how good a result Newton's rules may yield.

On this subject, d'Alembert wrote, in his *Recherches sur differens points importans du système du monde*, première partie (Paris, 1754—facsimile reprint: Culture et Civilization, Brussels, 1966), p. 203: 'On lit à la tête de la Théorie Newtonienne de la Lune qui se trouve dans le second volume de l'Astronomie de Gregory, que les Tables calculées d'après cette Théorie, diffèrent très-rarement de 2′ d'avec les observations, & que cet accord est confirmé par un grand nombre d'observations de M. *Flamsteed*. Cependant on a reconnu depuis que la différence alloit quelquefois beaucoup plus loin. Mais il a fallu un long temps pour s'en assurer. Qu'on juge sur cet exemple combien nous devons être circonspects dans nos assertions, quand même nous aurions en apparence un grand nombre d'observations conformes à nos nouvelles Tables.'

44. At the December 1973 meeting of the History of Science Society, in San Francisco, Mr Waff (see n. 38 *supra*) presented a report on 'The Lunar Theory of Isaac Newton', of which he kindly made a copy available to me. He concludes: Newton's 'qualitative explanations of the physical causes of the principal lunar inequalities were a favorite textbook demonstration of the usefulness and validity of his philosophy. His quantitative determinations were perhaps performed with the aid of too many simplifying assumptions, so that although his computations were more accurate than previous empirical estimates, his methods were quickly dropped when greater accuracy became desirable and more powerful analytical techniques were developed. Yet Newton's determinations were the first of their kind, and a measure of the reputation which he must have gained from them can be seen from the fact that nearly all *new* lunar tables constructed during the first half of the eighteenth century utilized in some fashion his tabular theory. During the half century or so after the publication of the *Principia*, "lunar theory", considered in all three ways [qualitative explanations, quantitative determinations, and lunar tables] I have described, moved from a solely astronomical subject to one which provided a touchstone for the acceptance of Newton's inverse-square law of gravitation. For this reason, it appears to me not to have deserved the neglect which historians have hitherto ascribed to it. On the contrary, a consideration of it seems to me essential for any study of Newton's influence generally during the eighteenth century'.

45. For instance, Richard Stevenson's *Newton's Lunar Theory, Exhibited Analytically* (Cambridge: J. & J. J. Deighton; London: Whittaker, Treacher & Arnot, 1834) is a very useful book, but in fact it presents a composite or synthesized and transformed 'Newtonian' theory rather than a 'Newton's lunar theory'.

46. The published researches of Eric Forbes and Victor E. Thoren have begun to change the traditional views on the fate of Newton's theory of the moon and its influence, and the lunar theory of Newton's predecessors. The forthcoming volume of Newton's *Correspondence*, edited by A. Rupert Hall, contains important source-materials and

editorial analyses of great value to all students of Newton's lunar theory. And the dissertation being completed by Craig Waff will certainly illuminate the post-Newtonian lunar theory as well as the work of Newton himself.

It is rather astonishing to observe how much progress has been made (and is currently being made) in this aspect of history in recent years, and how little was done in the century or more since Baily. All the more reason to hope for a competent and full analysis of Newton's own work on the moon.

A major contribution to certain aspects of Newton's lunar theory is given in vol. 6 of D. T. Whiteside's edition of *The Mathematical Papers of Isaac Newton*. The documents there printed, together with Whiteside's analysis, contradict Rouse Ball's assertions concerning Newton's statement (see note 2 *supra*) that 'Apsis lunæ est duplo velocior circiter'.

It would be most useful to have a careful analysis of Newton's attempts to produce a satisfactory lunar theory (in the 1690's), and the stages whereby he either partially or totally abandoned the program of deriving such a theory by mathematical methods applied to gravitational celestial mechanics. The documents for such a study are available primarily in MS Add. 3966. In particular, such an inquiry would reveal how completely Newton may have abandoned the mathematical-gravitational methods at the time of writing the 'rules' printed in the *Theory of the Moon's Motion* (1702), and whether the alterations introduced when these 'rules' were incorporated in the new scholium to Prop. 35, Book III (ed. 2) were based solely on an epicyclic model in which Flamsteed's observations were used or had any input from mathematical-gravitational methods, as Newton himself alleged.

7. A Supplemental Enquiry: Was Edmond Halley the Editor of the *Miscellanea Curiosa*?

The suggested association of Halley with Newton's *Theory of the Moon's Motion* (1702) was based on the fact that this published English version of Newton's essay was printed again in the *Miscellanea Curiosa*, of which Halley is supposed to have been the editor. Augustus De Morgan, who originally made the suggestion (see Introduction, § 4), did not know that there had been an earlier printing of the version of Newton's essay in Harris's *Lexicon*, a fact which greatly weakens his argument; furthermore, De Morgan assumed without question that Halley was the editor of the *Miscellanea*. But was he? I have found no firm evidence to this effect: no letter written by Halley or addressed to him, in which specific mention is made of his having been associated in any way with this publication.[1]

Until fairly recently, the *Miscellanea Curiosa* did not regularly appear in booksellers' catalogues as a production of Halley's.[2] In the *Term Catalogues, 1662–1709* (*1711*), ed. by Edward Arber (3 vols.: London, 1903–1906), the *Miscellanea Curiosa* appears without the name of a compiler or editor; but in the index (which was made for Arber's reprint), Halley's name occurs as compiler. No evidence is given for this attribution, and we may suspect that the information was derived from the catalogue of the British Museum, where Halley is named as the compiler without even the benefit of a question-mark.

The attribution to Halley of a major role in the production of the *Miscellanea* began in his own times. The earliest explicit statement that Halley was responsible for the *Miscellanea*[3] occurs in the first published biography of Halley, in the *Biographia Britannica*, vol. 4 (London, 1757), p. 2514: 'In the midst of all this business came out the *Miscellanea Curiosa*, containing besides others, several of his pieces, and the whole printed under his direction in 1708, in three volumes, 8vo.'[4] This clear reference to the work having been 'printed under his direction' may carry authority since it was published only some fifteen years after Halley's death. Furthermore, this biography is said to have been written with the assistance of Halley's son-in-law, who would presumably have had access to Halley's papers and possibly to annotated copies of his publications.[5]

Halley must have had some association with the production of the *Miscellanea*, even if he was not the actual editor-compiler, since some of his articles appear in a version which is revised or altered from their original printing in the *Philosophical Transactions*.[6] Furthermore, the first six articles in vol. 1 are by Halley, and he proves to be the author of more than half of the articles (13 out of 23) making up this volume, which implies that if the compiler was not Halley himself, he must have been a great admirer of Halley's. Additionally, the man who put the volume together would have had to have been someone in contact with Newton, in order to have obtained from him the errata and emendata to his essay on the moon's motion, which was published in vol. 1 of the *Miscellanea*; Halley has this qualification.

The foregoing information does not constitute absolutely convincing evidence that Halley was the actual editor or compiler of the *Miscellanea*. For even if Halley had been closely associated with the production of this collection, his name does not appear on the title-page. Furthermore, Halley is not mentioned in the preface itself, nor is he named as author of the preface.[7] This preface is written in a rather pompous manner, which appears to be very different from his characteristic style;[8] he could have been compiler or editor, of course, even though the preface was written by someone else.[9] The fact that the *Miscellanea* was not reviewed in the *Philosophical Transactions* may also be of significance, since we would have expected that some notice would have been taken in those pages of a work by Halley. The third edition of the *Miscellanea* was issued as '*Revised and Corrected*, by W. Derham, F.R.S.' The introduction of Derham's name, without that of the original editor, would support the suspicion that the original editor may not have been Halley; would not Halley have at last acknowledged, as his own, a work so successful as to require a third edition?

The *Miscellanea* is not listed among Halley's publications in either of the bibliographies published by Eugene Fairfield MacPike: *Correspondence and Papers of Edmond Halley* (Oxford: at the Clarendon Press, 1932), Appendix XIX, 'List of Halley's Published Writings'; *Dr. Edmond Halley (1656-1742). A Bibliographical Guide to his Life and Work arranged Chronologically* (London, 1939). I would not, however, give much weight to this omission, since the *Miscellanea* should have been included in any case—whether MacPike supposed it to have been edited by Halley or not—simply on the grounds that it contains so many of Halley's contributions (some of them in a slightly revised or altered form).

The sentence quoted earlier about Halley and the *Miscellanea* (taken from the *Biographia Britannica*, 1757) was reprinted with some variations by Benjamin Martin, in his *Biographia Philosophica* (London, 1764), p. 439, as follows: 'Very soon after he had finished this Business, to the great Satisfaction of the Learned, in 1708, he published the *Miscellanea Curiosa*, containing several of his own, as well as many curious Discoveries and Observations of others, which had come under his Notice.' In a bibliography of Halley, published in R. Watt's *Bibliotheca Britannica*, vol. 1 (London, 1824), col. 460e, an entry reads: 'Miscellanea Curiosa. 1708, 3 vols. 8vo. This was published under his direction.'[10]

Colin A. Ronan: *Edmond Halley, Genius in Eclipse* (Garden City, New York: Doubleday & Company, 1969), p. 189, says that Halley 'edited three volumes called *Miscellanea Curiosa* . . . which was a selection of the more interesting papers that had appeared in the *Philosophical Transactions* in the hope that they would reach a wider audience of "inquisitive Gentlemen".' This statement does not appear to have been based on a detailed examination of either the publication itself or the circumstances of its production. Angus Amitage's *Edmond Halley* (see n. 6 *infra*), contains the following statement, on pp. 156–157, for which no supporting evidence is given:

> Soon after his appointment to the Savilian Chair, Halley published a *réchauffé* of some of the most interesting papers so far contributed to the Royal Society, hoping that they would thereby reach a wider circle of readers and remain no longer 'so obscurely hid, that but very few inquisitive Gentlemen ever so much as heard of them'. This was the origin of *Miscellanea Curiosa, being a Collection of some of the Principal Phenomena in Nature*, etc. (3 vols., London, 1705–7). Halley's Preface (his name does not appear) suggests a limitation to papers of mathematical and physical interest which had already passed 'the Censure of the Learned World,' and the collection includes, among other items, the cream of his own contributions, with a world chart showing both magnetic variation and wind distribution. However, the selection embraces memoirs of biological and geographical content as well.

Hence, the conclusion to which we are led is only that Halley was in some way associated with the production of the *Miscellanea*, if only to the minimal extent of providing revised versions of his own contributions. If he had been the editor (or compiler), we could then easily account for the fact that the first volume opens with his papers and contains so high a proportion of papers written by him; for surely, in 1708, not many men of that time who were learned in science would have given Halley and his contributions such great prominence. Halley as editor would easily have obtained from Newton the errata and emendata that appear in volume 1. No doubt, a thorough search through the MS Council Minutes of the Royal Society or the Journal Book might reveal evidence of the editorship of the *Miscellanea*, for it is difficult to imagine such a collection being put together and published without the permission of the Royal Society; at any rate, these volumes would surely have come to the attention of the officers and members of the council after publication. Some years before the *Miscellanea* issued from the press, i.e., on 7 December 1692, the Council Minutes do contain this information: 'Halley offered that if it shall be undertaken to print a book of Philosophicall matters such as the Transactions used to consist of, that he would undertake to furnish de proprio five sheets in twenty.' May it not therefore appear that Halley had been planning to produce just such a collection as the *Miscellanea Curiosa*?

We may conclude that there seems little reason to contradict the two-centuries-old tradition of Halley's editorship. Furthermore, this examination of Halley in relation to the *Miscellanea* does not help us to identify the translator of Newton's *Theory of the Moon's Motion* (1702), nor the author of the Preface to that work.

Notes

1. Nor have I found any statement printed during Halley's life-time to indicate that he was the editor (or compiler) of the *Miscellanea*. In none of the three editions of the *Miscellanea* is there even a hint as to the identity of the editor, or the author of the general Preface (to vol. 1).

I hope that any reader who may have further information on the subject will make it speedily available to the scholarly public.

2. For example, Halley's editorship is not mentioned in the description of the *Miscellanea* in Sotheran's *Bibliotheca Chemico-Mathematica*, compiled and annotated by H.Z. [=Heinrich Zeitlinger] and H.C.S. [=Henry Cecil Sotheran], (London: Henry Sotheran and Co., 1921), vol. 1, # 3011; nor in the First Supplement (London: Henry Sotheran, 1932), # 475, # 478, nor in the Third Supplement (London: Henry Sotheran, 1952), # 401, # 402.

3. I am grateful to Sir Edward C. Bullard of Cambridge University for bringing this fact to my attention, and for general advice concerning Halley and the *Miscellanea*.

4. The 'business' in question was Halley's mission to help 'to fix on a proper Place for making a commodious Harbour on the Adriatic'; he had also just finished preparing for the press an edition of the

'Spherics' of Menelaus and a translation 'of *Apollonius* from the *Arabic* into Latin'.

5. It is stated by the biographer that he had access to some manuscripts made available to him by Henry Price, Halley's son-in-law. A manuscript note concerning this fact, written by S. P. Rigaud, is printed in Eugene Fairfield MacPike: *Correspondence and Papers of Edmond Halley* (Oxford: at the Clarendon Press, 1932), p. v. See, further, S. P. Rigaud: 'Some Particulars of the Life of Dr. Halley . . .', *Monthly Notices of the Royal Astronomical Society*, 12 Dec. 1834, vol. 3, p. 67.

6. Sir Edward C. Bullard (personal communication; 24 Sept. 1971) writes 'that the wind map facing p. 80 is different from the one in *Phil. Trans.* Vol. 16 no. 183 facing p. 151.' And Angus Armitage: *Edmond Halley* (London: Nelson, 1966) has observed (p. 131) that when 'Halley's paper on the Breslau Bills of Mortality was reprinted [from the *Philosophical Transactions*] in his [!] *Miscellanea Curiosa* of 1705–7', he added to 'the original text . . . some further considerations bearing upon current social problems and reflecting his attitude towards these.'

7. Some doubt arises concerning Halley's authorship of the preface, since it is there said that all the papers in the volume are mere reprints, whereas Halley's own papers are revised, as pointed out in n. 6 *supra*. See, further, n. 8 *infra*.

8. Sir Edward C. Bullard, to whom I am indebted for bio-bibliographical information concerning Halley and the *Miscellanea*, informs me (see n. 6 *supra*) that he finds this preface 'rather pompous' and concludes, 'I doubt if it was written by him.'

9. Even if Halley had been the compiler of the *Miscellanea*, he might have withdrawn his name on finding the work described on the title-page as a 'COLLECTION of some of the Principal *PHÆNOMENA* in NATURE' which are 'Accounted for by the Greatest Philosophers of this Age', the 'Most Valuable DISCOURSES, Read and Delivered to the *ROYAL SOCIETY*'. For how could Halley have subscribed to this publisher's blurb in relation to a volume in which he appeared not only as the most prominent contributor, but the author of more than half of the contents!

10. It may be observed that Martin, and—following Martin—Watt, used the date of the second edition, 1708, rather than 1705, when vol. 1 of the *Miscellanea* was first published.

Sir Edward C. Bullard has written (personal communication to the writer, 24 Sept. 1971) that Watt may not be too reliable an authority, since 'he includes various things that are unlikely to be Halley's (e.g., the *Declaration of the Heavens and of the Earthly Flat Form*).'

8. Newton's Corrections to the Text of the *Theory of the Moon's Motion* (1702)

Following the table of contents to vol. 1 of the *Miscellanea Curiosa* (London, 1705), there appear a set of 'Corrections made by Mr. Isaac Newton to the Theory of the Moon'. These corrections are reprinted here, but the references to page, paragraph and line have been altered so as to apply to the pamphlet, the *Theory of the Moon's Motion* (1702):

Page 10, par. 2, line 9, read shall scarce be above two Minutes in her Syzygys, or above three in her Quadratures.

Page 11, par. 2, line 2, for 35′ *read* 45′

Page 12, par. 2, line 2, for 247 *read* 267 *and in par. 3, line 6, for* 0″ *read* 40″ *and in line 11, for* 247 *read* 267

Page 17, par. 2, line 8, for compared *read* computed

II. A NEW AND MOST ACCURATE THEORY OF THE MOON'S MOTION ... Written by That Incomparable Mathematician Mr. Isaac Newton (London, 1702)

A New and moſt Accurate

THEORY

OF THE

Moon's Motion;

Whereby all her Irregularities may be ſolved, and her Place truly calculated to Two Minutes.

Written by
That Incomparable Mathematician
Mr. Isaac Newton,

And Publiſhed in *Latin* by
Mr. David Gregory
in his *Excellent Aſtronomy.*

LONDON,
Printed, and ſold by *A. Baldwin* in
Warwick-lane. 1702.

TO THE
READER.

THE Irregularity of the Moon's Motion hath been all along the just Complaint of Astronomers; and indeed I have always look'd upon it as a great Misfortune that a Planet so near us as the Moon is, and which might be so wonderfully useful

To the Reader.

ful to us by her Motion, as well as her Light and Attraction (by which our Tides are chiefly occasioned) should have her Orbit so unaccountably various, that it is in a manner vain to depend on any Calculation of an Eclipse, a Transit, or an Appulse of her, tho never so accurately made. Whereas could her Place be but truly calculated, the Longitudes of Places would be found every where at Land with great Facility, and might be nearly guess'd at Sea without the help of a Telescope, which cannot there be used.

This Irregularity of the Moon's Motion depends (as is now well known, since Mr. Newton *hath demonstrated the Law of Universal Gravitation) on the Attraction of*

To the Reader.

of the Sun, which perturbs the Motion of the Moon (and of all other Satellites *or secondary Planets) and makes her move sometimes faster and sometimes slower in her Orbit; and makes consequently an Alteration in the Figure of that Orbit, as well as of its Inclination to the Plain of the Ecliptick. But this being now to be accounted for, and reduced to a Rule; by this Theory such Allowances are made for it, as that the Place of the Planet shall be truly Equated.*

This therefore being perfectly New, and what the Lovers of Astronomy have a long while been put in hopes to receive from the Great Hand *that hath now finished*

To the Reader.

shed it; I thought it would be of good service to our Nation to give it an English *Dress, and publish it by it self: For as* Dr. Gregory's Astronomy *is a large and scarce Book, it is neither every ones Money that can purchase it, nor Acquaintance that can procure it; and besides I hope we have a great many Persons in* England *that have Skill and Patience enough to calculate a Planet's Place, who yet it may be don't well enough understand the* Latin Tongue *to make themselves Masters of this Theory in the Author's own Words. At least I perswade my self, that a Theory so easy and plain as this, which carries along with it such a Pretence to Exactness, will encourage many Persons to imploy themselves in Astronomical*

nomical Calculation, which before possibly they neglected, because they judged there was but little Exactness to be attained in it. And this would be a very useful way of spending their leisure Hours; and if they would oblige us with the Publication of good Ephemerides, Tables, &c. they would soon enflame others with a Desire of pursuing these kind of Studies.

The

The Famous Mr. *Isaac Newton*'s THEORY OF THE MOON.

THIS Theory hath been long expected by the Lovers of Art, and is now publish'd in Dr. *Gregory*'s Astronomy, in Mr. *Newton*'s own Words.

B By

By this Theory, what by all Astronomers was thought most difficult and almost impossible to be done, the Excellent Mr. *Newton* hath now effected, *viz.* to determine the Moon's Place even in her Quadratures, and all other Parts of her Orbit, besides the Syzygys, so accurately by Calculation, that the Difference between that and her true Place in the Heavens shall scarce be two Minutes, and is usually so small, that it may well enough be reckon'd only as a Defect in the Observation. And this Mr. *Newton* experienced by comparing it with very many Places of the Moon observ'd by Mr. *Flamstead,* and communicated to him.

The Royal Observatory at *Greenwich* is to the West of the Meridian of *Paris* 2°. 19′. Of *Uraniburgh* 12°. 51′. 30″. And of *Gedanum* 18°. 48′.

The

The mean Motions of the Sun and Moon, accounted from the Vernal Equinox at the Meridian of *Greenwich*, I make to be as followeth. The laſt Day of *December* 1680. at Noon (Old Stile) the mean Motion of the Sun was 9 Sign. 20°. 34′. 46″. Of the Sun's Apogæum was 3 Sign. 7°. 23′. 30″.

The mean Motion of the Moon at that time was 6 Sign. 1°. 35′. 45″. And of her Apogee 8 Sign. 4°. 28′. 5″. Of the Aſcending Node of the Moon's Orbit 5 Sign. 24°. 14′. 35″.

And on the laſt Day of *December* 1700. at Noon, the mean Motion of the Sun was 9 Sign. 20°. 43′. 50″. Of the Sun's Apogee 3 Sign. 7°. 44′. 30″. The mean Motion of the Moon was 10 Sign. 15°. 19′. 50″. Of the Moon's Apogee 11 Sign. 8°. 18′. 20″. And of her aſcending Node 4 Sign. 27°. 24′. 20″. For in 20 *Julian* Years or 7305 Days, the Sun's Motion is 20 Revolut. 0 Sign. 0°. 9′. 4″. And the Motion of the Sun's Apogee 21′. 0″.

The Motion of the Moon in the same Time is 247 Rev. 4 Sign. 13°. 34′. 5″. And the Motion of the Lunar Apogee is 2 Revol. 3 Sign. 3°. 50′. 15″. And the Motion of her Node 1 Revol. 0 Sign. 26°. 50′. 15″.

All which Motions are accounted from the Vernal Equinox: Wherefore if from them there be subtracted the Recession or Motion of the Equinoctial Point *in Antecedentia* during that space, which is 16′. 0″. there will remain the Motions in reference to the Fixt Stars in 20 *Julian* Years; *viz.* the Sun's 19 Revol. 11 Sign. 29°. 52′. 24″. Of his Apogee 4′. 20″. And the Moon's 247 Revol. 4 Sign. 13°. 17′. 25″. Of her Apogee 2 Revol. 3 Sign. 3°. 33′. 35″. And of the Node of the Moon 1 Revol. 0 Sign. 27°. 6′. 55″.

Accord-

(13)

According to this Computation the *Tropical Year* is 365 Days. 5 Hours. 48'. 57''. And the *Sydereal Year* is 365 Days. 6 Hours. 9'. 14''.

These mean Motions of the Luminaries are affected with various Inequalities: Of which,

1. There are the Annual Equations of the aforesaid mean Motions of the Sun and Moon, and of the Apogee and Node of the Moon.

The Annual Equation of the mean Motion of the Sun depends on the Eccentricity of the Earth's Orbit round the Sun, which is $16\frac{1}{12}$ of such Parts, as that the Earth's mean Distance from the Sun shall be 1000: Whence 'tis called the *Equation of the Centre*; and is when greatest 1°. 56'. 20''.

The greatest Annual Equation of the Moon's mean Motion is 11'. 49''. of her Apogee 20'. and of her Node 9'. 30''.

And

And these four Annual Equations are always mutually proportional one to another: Wherefore when any of them is at the greatest, the other three will also be greatest; and when any one lessens, the other three will also be diminished in the same *Ratio*.

The Annual Equation of the Sun's Centre being given, the three other corresponding Annual Equations will be also given; and therefore a Table of *that* will serve for all. For if the Annual Equation of the Sun's Centre be taken from thence, for any Time, and be called P, and let $\frac{1}{10}P = Q$, $Q + \frac{1}{60}Q = R$, $\frac{1}{6}P = D$, $D + \frac{1}{30}D = E$, and $D - \frac{1}{60}D = 2F$; then shall the Annual Equation of the Moon's mean Motion for that time be R, that of the Apogee of the Moon will be E, and that of the Node F.

Only observe here, that if the Equation of the Sun's Centre be required

quired to be *added*; then the Equation of the Moon's mean Motion muſt be *ſubtracted*, that of her Apogee muſt be *added*, and that of the Node *ſubducted*. And on the contrary, if the Equation of the Sun's Centre were to be *ſubducted*, the Moon's Equation muſt be *added*, the Equation of her Apogee *ſubducted*, and that of her Node *added*.

There is alſo an *Equation of the Moon's mean Motion* depending on the Situation of her Apogee in reſpect of the Sun; which is *greateſt* when the Moon's Apogee is in an Octant with the Sun, and is nothing at all when it is in the Quadratures or Syzygys. This Equation, when greateſt, and the Sun *in Perigæo*, is $3'. 56''$. But if the Sun be *in Apogæo*, it will never be above $3'. 34''$. At other Diſtances of the Sun from the Earth, this Equation, when greateſt, is reciprocally as the Cube of ſuch Diſtance. But when the

the Moon's Apogee is any where but in the *Octants*, this Equation grows less, and is mostly at the same distance between the Earth and Sun, as the Sine of the double Distance of the Moon's Apogee from the next Quadrature or Syzygy, to the Radius.

This is to be *added* to the Moon's Motion, while her Apogee passes from a Quadrature with the Sun to a Syzygy; but is to be *subtracted* from it, while the Apogee moves from the Syzygy to the Quadrature.

There is moreover another *Equation of the Moon's Motion*, which depends on the Aspect of the Nodes of the Moon's Orbit with the Sun: and this is *greatest* when her Nodes are in *Octants* to the Sun, and vanishes quite, when they come to their Quadratures or Syzygys. This Equation is proportional to the Sine of the double Distance of the Node

Node from the next Syzygy or Quadrature; and at greatest is but 47″. This must be *added* to the Moon's mean Motion, while the Nodes are passing from their Syzygys with the Sun to their Quadratures with him; but *subtracted* while they pass from the Quadratures to the Syzygys.

From the Sun's true Place take the equated mean Motion of the Lunar Apogee, as was above shewed, the Remainder will be the Annual Argument of the said Apogee. From whence the *Eccentricity of the Moon,* and the *second Equation* of her Apogee may be compar'd after the manner following *(which takes place also in the Computation of any other intermediate Equations.)*

(18)

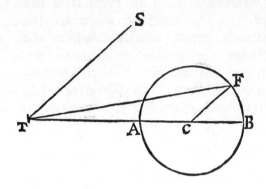

Let T reprefent the Earth, T S a Right Line joining the Earth and Sun, T A C B a Right Line drawn from the Earth to the middle or mean Place of the Moon's Apogee, equated, as above: Let the Angle S T A be the Annual Argument of the aforefaid Apogee, T A the leaft Eccentricity of the Moon's Orbit, T B the greateft. Bifect A B in C; and on the Centre C with the Diftance A C defcribe a Circle A F B, and make the Angle B C F = to the
double

double of the Annual Argument. Draw the Right Line T F, that shall be the Eccentricity of the Moon's Orbit; and the Angle B T F is the second Equation of the Moon's Apogee required.

In order to whose Determination let the mean Distance of the Earth from the Moon, or the Semidiameter of the Moon's Orbit, be 1000000; then shall its *greatest Eccentricity* T B be 66782 such Parts; and the *least* T A, 43319. So that the greatest Equation of the Orbit, *viz.* when the Apogee is in the Syzygys, will be 7°. 39'. 30". or perhaps 7°. 40'. (for I suspect there will be some Alteration according to the Position of the Apogee in ♋ or in ♑.) But when it is in Quadrature to the Sun, the greatest Equation aforesaid will be 4°. 57'. 56". and the greatest Equation of the Apogee 12°. 15'. 4".

Having from these Principles made a Table of the Equation of the Moon's Apogee, and of the Eccentricitys of her Orbit to each degree of the Annual Argument, from whence the Eccentricity T F, and the Angle B T F (*viz.* the second and principal Equation of the Apogee) may easily be had for any Time required; let the Equation thus found be added, to the first Equated Place of the Moon's Apogee, if the Annual Argument be less than 90°, or greatear than 180°, and less than 270; otherwise it must be subducted from it: and the Sum or Difference shall be the Place of the Lunar Apogee secondarily equated; which being taken from the Moon's Place equated a third time, shall leave the mean Anomaly of the Moon corresponding to any given Time. Moreover, from this mean Anomaly of the Moon, and the before-found Eccentricity of her Orbit, may be found (by means of a Table of Equations

quations of the Moon's Centre made to every degree of the mean Anomaly, and fome Eccentricitys, *viz.* 45000, 50000, 55000, 60000, and 65000) the *Proſtaphæreſis* or Equation of the Moon's Centre, as in the common way: and this being taken from the former Semicircle of the middle Anomaly, and *added* in the latter to the Moon's Place thus thrice equated, will produce the Place of the Moon a fourth time equated.

The greateſt Variation of the Moon (*viz.* that which happens when the Moon is in an Octant with the Sun) is, nearly, reciprocally as the Cube of the Diſtance of the Sun from the Earth. Let that be taken $37'. 25''$. when the Sun is *in Perigæo*, and $33'. 40''$. when he is *in Apogæo:* And let the Differences of this Variation in the Octants be made reciprocally as the Cubes of the Diſtances of the Sun from the Earth; and ſo let a Table be be made of the aforesaid Variation of the Moon in her *Octants* (or its Logarithms) to every *Tenth*, *Sixth*, or *Fifth* Degree of the mean Anomaly: And for the Variation out of the Octants, make, as Radius to the Sine of the double Distance of the Moon from the next Syzygy or Quadrature :: so let the aforefound Variation in the Octant be to the Variation congruous to any other Aspect; and this *added* to the Moon's Place before found in the first and third Quadrant (accounting from the Sun) or *subducted* from it in the second and fourth, will give the Moon's Place equated a fifth time.

Again, as Radius to the Sine of the Sum of the Distances of the Moon from the Sun, and of her Apogee from the Sun's Apogee (or the Sine of the Excess of that Sum above $360°$.) :: so is $2'.10''$. to a sixth Equation of the Moon's Place, which must be *subtracted*, if the

the aforesaid Sum or Excess be less than a Semicircle, but *added*, if it be greater. Let it be made also, as Radius to the Sine of the Moon's Distance from the Sun : : so 2′. 20″. to a seventh Equation: which, when the Moon's Light is encreasing, *add*, but when decreasing, *subtract*; and the Moon's Place will be equated a seventh time, and this is her Place *in her proper Orbit*.

Note here, the Equation thus produced by the mean Quantity 2′. 20″. is not always of the same Magnitude, but is encreased and diminished according to the Position of the Lunar Apogee. For if the Moon's Apogee be in Conjunction with the Sun's, the aforesaid Equation is about 54″. greater: but when the Apogees are in opposition, 'tis about as much less; and it librates between its greatest Quantity 3′. 14″. and its least 1′. 26″. And this is when the Lunar Apogee is in Conjunction or Opposition with the

the Sun's: But in the Quadratures the aforesaid Equation is to be lessen'd about 50″. or one Minute, when the Apogees of the Sun and Moon are in Conjunction; but if they are in Opposition, for want of a sufficient number of Observations, I cannot determine whether it is to be lessen'd or increas'd. And even as to the Argument or Decrement of the Equation 2′. 20″. above mentioned, I dare determine nothing certain, for the same Reason, *viz.* the want of Observation accurately made.

If the sixth and seventh Equations are augmented or diminished in a reciprocal *Ratio* of the Distance of the Moon from the Earth, *i. e.* in a direct *Ratio* of the Moon's Horizontal Parallax; they will become more accurate: And this may readily be done, if Tables are first made to each Minute of the said Parallax, and to every sixth or fifth Degree of the Argumennt of the sixth fixth Equation for the *Sixth*, as of the Distance of the Moon from the Sun, for the *Seventh* Equation.

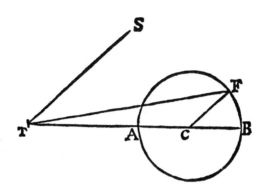

From the Sun's Place take the mean Motion of the Moon's ascending Node, equated as above; the Remainder shall be the Annual Argument of the Node, whence its second Equation may be computed after the following manner in the preceding Figure:

Let

Let T as before represent the Earth, TS a Right Line conjoining the Earth and Sun: Let also the Line TACB be drawn to the Place of the Ascending Node of the Moon, as above equated; and let STA be the Annual Argument of the Node. Take TA from a Scale, and let it be to AB :: as 56 to 3, or as $18\frac{1}{3}$ to 1. Then bisect BA in C, and on C as a Centre, with the Distance CA, describe a Circle as AFB, and make the Angle BCF equal to double the Annual Argument of the Node before found: So shall the Angle BTF be the second Equation of the Ascending Node: which must be *added* when the Node is passing from a Quadrature to a Syzygy with the Sun, and *subducted* when the Node moves from a Syzygy towards a Quadrature. By which means the true Place of the Node of the Lunar Orbit will be gained: whence from Tables made after the common way, the

Moon's

Moon's Latitude, and the Reduction of her Orbit to the Ecliptick, may be computed, supposing the Inclination of the Moon's Orbit to the Ecliptick to be 4°. 59′. 35″. when the Nodes are in Quadrature with the Sun; and 5°. 17′. 20″. when they are in the Syzygys.

And from the Longitude and Latitude thus found, and the given Obliquity of the Ecliptick 23°. 29′. the Right Ascension and Declination of the Moon will be found.

The Horizontal Parallax of the Moon, when she is in the Syzygys at a mean Distance from the Earth, I make to be 57′. 30″. and her Horary Motion 33′. 32″. 32‴. and her apparent Diameter 31′. 30″. But in her Quadratures at a mean Distance from the Earth, I make the Horizontal Parallax of the Moon to be 56′. 40″. her Horary Motion 32′. 12″. 2‴. and her apparent Diameter 31′. 3″.

3″. The Moon in an Octant to the Sun, and at a mean Distance, hath her Centre distant from the Centre of the Earth about $60\frac{2}{3}$. of the Earth's Semi-Diameters.

The Sun's Horizontal Parallax I make to be 10″. and its apparent Diameter at a mean Distance from the Earth, I make 32′. 15″.

The Atmosphere of the Earth, by dispersing and refracting the Sun's Light, casts a Shadow as if it were an Opake Body, at least to the height of 40 or 50 Geographical Miles (by a Geographical Mile I mean the sixtieth Part of a Degree of a great Circle, on the Earth's Surface) This Shadow falling upon the Moon in a Lunar Eclipse, makes the Earth's Shadow be the *larger or broader*. And to each Mile of the Earth's Atmosphere is correspondent a Second in the Moon's Disk, so that the Semidiameter of the Earth's Shadow projected upon

upon the Disk of the Moon is to be encreafed about 50 Seconds: or which is all one, in a Lunar Eclipfe, the Horizontal Parallax of the Moon is to be encreafed in the Ratio of about 70 to 69.

Thus far the Theory of this Incomparable Mathematician. And if we had many Places of the Moon *accurately obferved*, efpecially about her Quadratures, and thefe well compared with her Places at the fame time calculated according to this Theory; it would then appear whether there yet remain any other fenfible Equations, which when accounted for, might ferve to improve and enlarge this Theory.

Dr. Greg. Aftr. Elem. Phyf. & Geom. p. 336.

F I N I S.

III. LUNÆ THEORIA NEWTONIANA, an extract (pp. 332–336) from David Gregory's *Astronomiæ Physicæ & Geometricæ Elementa* (Oxford, 1702)

ASTRONOMIÆ
PHYSICÆ & GEOMETRICÆ
ELEMENTA.

Auctore DAVIDE GREGORIO M. D.
Astronomiæ Professore Saviliano OXONIÆ,
& Regalis Societatis Sodali.

OXONIÆ,
E THEATRO SHELDONIANO, An. Dom. MDCCII.

SCHOLIUM.

Libet Lunæ Theoriam à D. *Newtono* in praxi usurpatam hic adjungere, quâ difficillimum istud & ab Astronomis hactenus desperatum consecutus est celeberrimus Philosophus; nimirum ut Lunæ Locum, etiam extra Syzygias, & in ipsis cum Sole Quadraturis, Cœlo adeo consentientem ex calculo definiat, (sicuti per plurima Lunæ Loca à Cl. D. *Flamstedio* observata expertus est,) ut dissensus à Cœlo, etiam cum maximus, vix duo scrupula prima adæquet; plerumque adeo sit exiguus, ut observandi incertitudini sit jure imputandus. In hac Calculi forma, quam ipsis Auctoris verbis expressam Astronomis sistimus, non attingit omnes omnino Inæqualitates quarum causæ superius sunt explicatæ, nedum illas de quibus adhuc tantum est suspicio; sed omissis iis quas se invicem compensaturas novit, aliisque minoris efficaciæ, eas duntaxat Æquationibus & Tabulis coercet, quarum majores sunt vires & effectus sensibiliores.

Lunæ Theoria Newtoniana.

'Observatorium *Grenovicense* occidentalius est *Parisiensi* $2^{gr}.19'$,
'*Uraniburgo* $12^{gr}.51'.30''$ & *Gedano* $18^{gr}.48'$.

'Solis & Lunæ Motus medios ab Æquinoctio verno in Meridi-
'ano *Grenovicensi* pono sequentes: nempe *Anno* 1680 *Decembris*
'Die ultimo *Stylo Juliano* Meridie, Motus medius Solis est $9^{sig}.20^{gr}$.
'$34'.46''$. Apogæi Solis $3^{sig}.7^{gr}.23'.30''$. Motus medius Lunæ 6^{sig}.
'$1^{gr}.35'.45''$. Apogæi Lunæ $8^{sig}.4^{gr}.28'.5''$. Nodi ascendentis Or-
'bitæ Lunaris $5^{sig}.24^{gr}.14'.35''$. Et *Anno* 1700. *Decembris* Die
'ultimo *Stylo Juliano* Meridie, Motus medius Solis est $9^{sig}.20^{gr}.43'$.
'$50''$. Apogæi Solis $3^{sig}.7^{gr}.44'.30''$. Motus medius Lunæ 10^{sig}.
'$15^{gr}.19'.50''$. Apogæi Lunæ $11^{sig}.8^{gr}.18'.20''$. & Nodi ascendentis 4^{sig}.
'$27^{gr}.24'.20''$. Viginti enim Annis *Julianis*, sive diebus 7305, Solis
'motus est $20^{rev}.0^{sig}.0^{gr}.9'.4''$. Motus Apogæi Solis $21'.00''$. Lunæ
'Motus est $247^{rev}.4^{sig}.13^{gr}.34'.5''$. Apogæi Lunaris Motus $2^{rev}.3^{sig}$.
'$3^{gr}.50'.15''$. Nodi Motus $1^{rev}.0^{sig}.26^{gr}.50'.15''$. Omnes prædicti
'Motus sunt à puncto Æquinoctii verni. Quod si ab illis subduca-
'tur ipsius Æquinoctialis puncti Motus in antecedentia interea
'factus, sc. $16'.40''$; manebunt Motus respectu Fixarum in Annis
'20. *Julianis*, nempe Motus Solis $19^{rev}.11^{sig}.29^{gr}.52'.24''$. Apogæi
'Solis $4'.20''$. Lunæ $247^{rev}.4^{sig}.13^{gr}.17'.25''$. Apogæi Lunæ $2^{rev}.3^{sig}$.
'$3^{gr}.33'.35''$. Nodi Lunæ $1^{rev}.0^{sig}.27^{gr}.6'.55''$.

'Secundum hunc computum Annus Tropicus est $365^{dier}.5^{hor}.48'$.
'$57''$. Annusque Sydereus $365^{dier}.6^{hor}.9'.14\frac{1}{2}''$.

'Motus Medii Luminarium suprapositi variis afficiuntur Inæ-
'qualitatibus.

'Et primo sunt Annuæ Æquationes dictorum Motuum medio-
'rum Solis & Lunæ, & Apogæi Nodique Lunæ. Æquatio Annua
'Motûs medii Solis pendet ab Excentricitate Orbitæ Telluris circa
'Solem, quæ est partium $16\frac{11}{12}$, qualium mediocris Distantia Solis à
'Terra

‘Terra est 1000; indeque vocatur Æquatio Centri: estque, cum
‘maxima, 1ᵍʳ. 56′. 20″. Maxima Æquatio Annua Motus medii
‘Lunæ est 11′. 49″. Apogæi ejus 20′. Nodique 9′. 30″. Atque quatuor
‘istæ Æquationes Annuæ sunt semper sibi mutuo proportionales. Id-
‘eoque cum earum aliqua est maxima, tres reliquæ sunt etiam ma-
‘ximæ; diminutâ vero quâvis, minuuntur etiam & reliquæ in eadem
‘ratione : Unde datâ Æquatione Annuâ centri Solis, dantur & tres
‘reliquæ Annuæ Æquationes congruæ; illius igitur Tabula sufficit:
‘Nam si Æquatio Annua Centri Solis Tempori cuivis congrua inde
‘deprompta vocetur P, & fiat $\frac{1}{5}$ P = Q, Q + $\frac{1}{60}$ Q = R, $\frac{2}{3}$ P = D, D + $\frac{1}{60}$ D = E,
‘& D − $\frac{1}{60}$ D = 2 F; erit Æquatio Annua eidem Tempori congrua Lunæ
‘quidem R, Apogæi Lunaris E, & Nodi F. Adnotandum, si Æquatio
‘Centri Solis est addenda, Æquationem Lunæ prædictam esse sub-
‘ducendam, Apogæi Lunaris Æquationem addendam, Nodi vero
‘Æquationem subducendam; & è contra, si Æquatio Centri Solis
‘est subducenda, addenda erit Lunæ Æquatio, Apogæi vero subdu-
‘cenda, & Nodi addenda.

‘Alia est Æquatio Motûs medii Lunæ, pendens à situ Apogæi
‘Lunaris respectu Solis, quæ maxima est cum Apogæum Lunæ
‘versatur in Octante cum Sole, & nulla cum illud ad Syzygias vel
‘Quadraturas pervenerit. Æquatio hæc, quando maxima, ad 3′. 56″.
‘ascendit, Sole in Perigæo versante; si vero Sol Apogæum teneat,
‘non ultra 3′. 34″. In aliis Distantiis Solis à Terra Æquatio hæc
‘maxima est reciproce ut Cubus istius Distantiæ. At cum Lunæ
‘Apogæum est extra Octantes, Æquatio dicta evadit minor, estque
‘ad maximam, positâ eâdem distantiâ Terræ & Solis, ut sinus du-
‘plæ Distantiæ Lunaris Apogæi à proxima Syzygia vel Quadra-
‘tura ad radium. Additur hæc Motui Lunæ, dum Lunæ Apo-
‘gæum transit à Solis Quadrato ad Syzygiam; sed inde subducitur
‘in transitu Apogæi à Syzygia ad Quadratum.

‘Alia porro est Motûs Lunæ Æquatio, pendens ab Aspectu No-
‘dorum Orbitæ Lunaris cum Sole; estque maxima cum Nodi in
‘Solis Octantibus versantur, evanescitque cum hi ad Syzygias aut
‘Aspectum Quadratum appellunt. Æquatio hæc proportionalis est
‘sinui duplæ Distantiæ Nodi à proxima Syzygia aut Quadratura,
‘cumque maxima ad 47″ ascendit. Additur hæc Motui Lunæ, dum
‘Nodi transeunt à Solis Syzygiis ad ejusdem Quadraturas; & sub-
‘ducitur in eorum transitu à Quadraturis ad Syzygias.

‘A Solis Loco vero aufer Motum medium Apogæi Lunæ æ-
‘quatum, ut supra est ostensum; residuum erit Argumentum An-
‘nuum dicti Apogæi. Exinde computentur Lunæ Excentricitas &
‘secunda Æquatio ejus Apogæi modo sequenti, [*qui in aliis qui-
busvis intermediis Æquationibus computandis locum etiam habet:*]
‘Referat T Terram; TS rectam conjungentem Terram & Solem;
‘TACB rectam à Terra ductam ad Locum medium Apogæi Lunaris
‘ut supra æquatum; angulus STA Argumentum Annuum dicti A-
‘pogæi; TA Lunaris Orbitæ Excentricitatem minimam; TB ejus-
‘dem

'dem Excentricitatem maximam. Biseca AB in c, centro c per A
'describe circulum AFB; fiat angulus BCF æqualis duplo Argu-
'mento Annuo: juncta recta TF erit
'Lunaris Orbitæ Excentricitas, angu-
'lusque BTF erit secunda Apogæi
'Lunæ Æquatio. Ad horum determi-
'nationem sit mediocris Distantia
'Lunæ à Terra, sive Orbitæ Lunaris

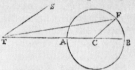

'Semidiameter, partium 1000000: Ejus maxima Excentricitas TB
'erit partium 66782, & minima TA earundem 43319; adeo ut
'maxima Orbis ejus Æquatio, cum sc. Apogæum est in Syzygiis,
'sit $7^{gr}.39'.30''$, vel forsan $7^{gr}.40'.00''$; (suspicio enim est hanc mu-
'tari pro situ Apogæi in ♋ vel ♑;) cum vero illud in Solis Quadrato
'hæret, dicta maxima Æquatio sit $4^{gr}.57'.56''$, & ut maxima Apo-
'gæi Æquatio sit $12^{gr}.15'.4''$.

'Constructâ ex hisce principiis Tabulâ Æquationum Apogæi
'Lunæ & Excentricitatum ejus Orbitæ ad singulos gradus Argu-
'menti Annui, unde Excentricitas TF & angulus BTF (sc. Æquatio
'secunda & præcipua Apogæi) dato Tempori congruentes facile pos-
'sint depromi; ad Locum Apogæi Lunæ primo æquatum ut supra
'addatur Æquatio modo inventa, si Argumentum Annuum minus
'sit 90^{gr}, aut majus 180^{gr}, minus vero quam 270^{gr}; secus vero ab eo
'subducatur: summa vel differentia erit Apogæi Lunaris Locus se-
'cundo æquatus; quo subducto ex Lunæ Loco tertio æquato, re-
'linquitur Lunæ Anomalia media dato Tempori congrua. Porro, ex
'hac Anomalia Lunæ media & modo inventa Orbis Excentricitate
'habebitur (ope Tabulæ Æquationum Centri Lunæ ad singulos
'Anomaliæ mediæ gradus, & aliquot Excentricitates v. g. 45000,
'50000, 55000, 60000 & 65000 fabricatæ) Prosthaphæresis sive Æ-
'quatio Centri Lunæ, ut vulgo; quâ subductâ in priori Anomaliæ
'mediæ semicirculo, additâ vero in posteriori ad Locum Lunæ
'hactenus ter æquatum, prodit Lunæ locus quarto æquatus.

'Maxima Lunæ Variatio, sc. quæ contingit cum Luna est in
'Octantibus Solis, est fere reciproce ut Cubus Distantiæ Solis à
'Terra. Capiatur ea $37'.25''$ cum Sol est Perigæo, & $33'.4''$ cum
'in Apogæo: fiantque Variationis hujus in Octantibus differentiæ
'reciproce ut differentiæ Cuborum Distantiarum Solis à Terra, &
'exinde construatur Tabula prædictæ Variationis Lunæ in Octanti-
'bus Solis (ejusve Logarithmi) ad singulos denos vel senos vel qui-
'nos gradus Anomaliæ mediæ: Et pro Variatione extra Octantes,
'fiat ut radius ad sinum duplæ Distantiæ Lunæ à proxima Syzygia
'vel Quadratura ita supra inventa Variatio in Octante ad Varia-
'tionem dato Aspectui congruam, quæ addita Loco Lunæ supra
'invento in primo & tertio quadrante, (computando à Sole,) aut ab
'eodem subducta in secundo & quarto, exhibet Lunæ Locum
'quinto æquatum.

'Rursus, ut radius ad sinum summæ Distantiarum Lunæ à Sole
&

Lib.IV. & Geometricæ Elementa. 335

'& Apogæi Lunæ ab Apogæo Solis (vel sinum excessus istius sum-
'mæ supra 360^{gr}) ita 2'. 10" ad sextam Loci Lunæ Æquationem,
'subducendam si prædicta summa vel dictus excessus minor sit semi-
'circulo, addendam si major.

'Fiat etiam ut radius ad sinum Distantiæ Lunæ à Sole ita 2'. 20"
'ad Æquationem septimam. Hanc aufer quando Lunæ Lumen au-
'getur, & (è contra) adde cum illud minuitur; & prodibit Lunæ Lo-
'cus septimo æquatus, quique est Locus ejus in propria Orbita.
'Notandum Æquationem, quæ hic effertur per mediocrem quan-
'titatem 2'. 20", non esse ejusdem semper magnitudinis, sed augeri
'& minui pro situ Lunaris Apogæi. Nam si Apogæum Lunare
'conjunctum fuerit cum Solis Apogæo, prædicta Æquatio est circi-
'ter 54" major; sin illi oppositum, tantundem minor: libratque
'inter maximam quantitatem 3'. 14" minimamque 1'. 26". Atque
'hæc obtinent ubi Apogæum Lunare est in Solis Syzygiis; ubi vero
'illud in Solis Quadrato hæret, minuenda est Æquatio prædicta 50
'circiter scrupulis secundis aut integro scrupulo primo, quando A-
'pogæum Lunæ & Solis Apogæum conjuncta sunt; si vero sunt
'opposita, propter Observationum penuriam affirmare nequeo au-
'gendane sit illa, an minuenda. Immo de supraposisitis incremento &
'decremento Æquationis 2'. 20", propter Observationum satis accu-
'ratarum defectum, certo statuere non ausim.

'Si Æquationes sexta & septima augeantur vel minuantur in ra-
'tione reciproca Distantiæ Lunæ à Terra, hoc est, in directa ratione
'Parallaxis horizontalis Lunæ; accuratiores fient. Atque istud
'prompte fiet, si prius Tabulæ fuerint constructæ ad singula scru-
'pula dictæ Parallaxis, singulosque senos vel quinos gradus cum Ar-
'gumenti Æquationis sextæ pro Æquatione sexta, tum Distantiæ
'Lunæ à Sole pro septima.

'A Loco Solis vero aufer medium Motum Nodi ascendentis
'Lunæ æquatum ut supra; residuum erit Nodi Argumentum
'Annuum; unde ejus Æquatio secunda computabitur modo se-
'quenti: In figura præcedente referat ut prius T Terram; TS re-
'ctam jungentem Terram & Solem: Referat porro TACB lineam
'ductam ad Locum Nodi ascendentis Lunæ ut supra æquatum, &
'STA Argumentum Annuum Nodi. Capiatur TA ad AB ut 56 ad
'3, sive 18⅔ ad 1. Bisecta BD in c, & centro c intervallo CA vel CB
'describe circulum AFB, fiatque angulus BCF æqualis duplo Argu-
'mento Nodi Annuo ut supra invento; eritque BTF angulus Æ-
'quatio secunda Nodi ascendentis, addenda in transitu Nodi à Solis
'Quadrato ad Syzygiam, subducenda in ejusdem transitu à Syzygia
'ad Quadraturam. Atque sic habetur Locus verus Nodi Orbitæ
'Lunaris: Unde ex Tabulis methodo vulgari constructis supputa-
'bitur Lunæ Latitudo & Reductio Lunæ ab Orbita sua ad Eclipti-
'cam, posita Inclinatione Orbis Lunaris ad planum Eclipticæ 4^{gr}. 59'.
'35" cum Nodi sunt in Solis Quadrato; & 5^{gr}. 17'. 20" cum iidem
'in Syzygiis versantur. Ex modo inventis Longitudine & Lati-
'tudine

'tudine & data Obliquitate Ecliticæ 23ʳ. 29', Lunæ Ascensio recta
'& Declinatio eruentur.

'Lunæ in Syzygiis mediocriter distantis à Terra Parallaxin ho-
'rizontalem pono 57'. 30"; Motum horarium 33'. 32". 32"'; & Dia-
'metrum apparentem 31'. 30": In Quadraturis vero mediocriter à
'Terra distantis Parallaxin pono 56'. 40"; Motum horarium 32'. 12".
'2"'; & Diametrum apparentem 31'. 3". Lunæ in Solis Octante me-
·diocriter distantis centrum distat à centro Terræ quasi 60½ Semi-
'diametrorum Terræ.

'Solis Parallaxin horizontalem pono 10"; & 32'. 15" apparentem
'ejus Diametrum in mediocri Distantia à Terra.

'Telluris Atmosphæra, refringendo & dissipando Solis Lumen,
'Umbram projicit, perinde ac si opaca foret, ad altitudinem mini-
'mùm 40 aut 50 milliarium Geographicorum: (Milliare Geogra-
'phicum appello partem sexagesimam Gradûs magni Circuli in Tel-
'luris superficie:) Umbra hæc in Eclipsi Lunari in Lunam incidens
'Telluris Umbram auctiorem reddit. Et singulis milliaribus At-
'mosphæræ Terrestris respondent singula scrupula secunda in Lunæ
'disco. Adeoque Umbræ Terrestris Semidiameter in Lunæ discum
'projecta augenda est 50 circiter secundis, aut (quod eodem reci-
'dit) in Eclipsi Lunari Lunæ Parallaxis horizontalis augenda est
'in ratione circiter 70 ad 69.

Si plura Lunæ Loca accuratissime observata (præsertim circa
Quadraturas) cum ejusdem Locis ad eadem tempora ex suprapo-
sita Theoria supputatis conferantur; patebit tandem num aliæ sint
Æquationes sensibiles, quibus Theoria hæc est amplificanda.

PROPOSITIO XXX.

Tempus mediæ Conjunctionis aut Oppositionis Solis & Lunæ proxime insequentis ad datum Tempus determinare.

Conjunctio media est cum Locus Solis medius idem est cum Loco Lunæ medio in Ecliptica: Oppositio vero media cum ille huic opponitur. Quæratur ex Tabulis Motûs medii Lunæ à Sole Distantia media Lunæ à Sole dato Tempori (ad medium reducto) competens, quæ si nulla sit, vel Signorum sex, tunc erit ipsa Conjunctio aut Oppositio media; sin minor, notetur defectus, (sive arcus quem Luna percurrere debet ut Solem denuo assequatur si Conjunctio media desideretur, vel ut à Sole sex Signis elongetur si Oppositio,) & ex Tabula Motûs medii Lunæ à Sole in diebus, horis horæque partibus colligatur Tempus requisitum ad istum arcum percurrendum; subducendo nempe semper ab arcu adhuc percurrendo arcum in Tabula repertum proxime minorem, Tempus quo subductus percurritur notando, Temporaque hæc in unam summam colligendo: hæc addita Tempori dato conficiet Tempus Syzygiæ mediæ quæsitum.

Quod si Tempora aliquot deinceps Syzygiarum mediarum quærantur, addatur Tempori supra invento Tempus revolutionis integræ

IV. SIR ISAAC NEWTON'S THEORY OF THE MOON, an extract (vol. 2, pp. 562–571) from David Gregory's *The Elements of Astronomy, Physical and Geometrical* (London, 1715)

THE
ELEMENTS
OF
Aſtronomy,
PHYSICAL and GEOMETRICAL.

By DAVID GREGORY M. D. *Savilian* Profeſſor of ASTRONOMY at *Oxford*, and Fellow of the Royal-Society.

Done into *Engliſh*, with Additions and Corrections.

To which is annex'd,
Dr. HALLEY's Synopſis of the Aſtronomy of Comets.

In Two Volumes.

Vol. II.

LONDON:
Printed for J. NICHOLSON at the *King's-Arms* in *Little-Britain*, and ſold by J. MORPHEW near *Stationers-Hall*. M DCC XV.

Price of the Two Vols. 12s. Bound.

with others: As for Example, they do not purely deduce from the Moon's Orbit given, the Æquation of the Excentric (or of the Moon's Center) but as it were confusedly mix it with the others; as with that part of the Inequality of the Variation that depends upon the time between the Quadratures, which arises from the Situation of the Moon in respect of its Apogæum; see *Schol. Prop.* 18. But as they made their Tables not from known Physical Causes and their Periods, but only by attending to Observations; it is no wonder if they did not rightly distinguish the Inequalities from one another and dispos'd them under foreign or less proper Titles.

SCHOLIUM.

I have thought fit to subjoyn the Theory of the Moon made use of by Sir *Isaac Newton*, by which this incomparable Philosopher has compass'd this extremely difficult Matter, hitherto despair'd of by Astronomers; namely, by Calculation to define the Moon's Place even out of the Syzygies, nay, in the Quadratures themselves so nicely agreeable to its Place in the Heavens (as he has experienc'd it by several of the Moon's Places observ'd by the ingenious Mr. *Flamstead*) as to differ from it (when the difference is the greatest) scarce above two Minutes in her Syzygies, or above three in her Quadratures; but commonly so little that it may well enough be reckon'd only as a Defect of the Observation. In this Calculation, which we give in the Words of the Author, he does not wholly mention all the Inequalities, whose Causes are above explain'd, nor those which are as yet only suspected; but omitting those which he knew wou'd take off one another, and others

of

Book IV. *of* ASTRONOMY. 563

of less Moment, he only confines those to Æquations and Tables, that have the greatest Force and produce the most sensible Effects.

Sir Isaac Newton's *Theory of the Moon.*

The Royal Observatory at *Greenwich* is to the West of the Meridian of *Paris* 2°. 19', of *Uraniburg* 12°. 51'. 30", and of *Dantzick* 18°. 48'.

I put down the Sun and Moon's mean motions from the Vernal Æquinox at the Meridian of *Greenwich* as follows: *viz.* the last Day of *December* 1680 *Old Style*, at Noon, the Sun's mean Motion was 9^{sign}. 20°. 34'. 46". That of the Sun's Apogæum 3^{sig}. 7°. 23'. 30". The Moon's mean motion 6^{sig}. 1°. 45. 45" That of the Moon's Apogæum 8^{sig}. 4°. 28'. 5". That of the ascending Node of the Moon's Orbit 5^{sig}. 24°. 14'. 35". And *December* the last 1700 *Old Style* at Noon, the Sun's mean motion was 9^{sig}. 20°. 43'. 50". That of the Sun's Apogæum 3^{sig}. 7°. 44'. 30". The mean motion of the Moon 10^{sig}. 15°. 19'. 50". Of the Moon's Apogæum 11^{sig}. 8°. 18'. 20". And of the ascending Node 4^{sig}. 27°. 24'. 20". For in Twenty *Julian* Years, or in 7305 Days, the Sun goes thro' 20^{rev}. 0^{sig}. 0°. 9'. 4". The motion of the Sun's Apogæum 21'. 00". The Moon's motion 267^{rev}. 4^{sig}. 13°. 34'. 5". The motion of the Moon's Apogæum 2^{rev}. 3^{sig}. 3°. 50'. 15". Of the Node 1^{rev}. 0^{sig}. 26°. 50'. 15". All the foresaid motions are from a Point of the Vernal Æquinox. And if from them be substracted the Precession, or Retrograde Motion of the Æquinoctial Point it self, which has mov'd in the mean time *in antecedentia, viz.* 16'. 40": The Motions will remain in respect of the Fix'd Stars

Stars in 20 *Julian* Years, namely the motion of the Sun, 19^{rev}. 11^{fig}. $29°$. $52'$. $24''$. That of the Sun's Apogæum $4'$ $20''$. Of the Moon 267^{rev}. 4^{fig}. $13°$. $17'$. $25''$. Of the Moon's Apogæum 2^{rev}. 3^{fig}. $3°$. $33'$. $35''$. Of the Moon's Node 1^{rev}. 0^{fig}. $27°$. $6'$. $55''$.

According to this Computation the *Tropical Year* is of 365^{days}. 5^{hours}. $48'$. $57''$. And the *Sydereal Year*, of 365^{days}. 6^{hours}. $9'$. $14\frac{1}{2}''$.

The mean motions of the Luminaries abovementioned have several Inequalities.

First there are the Annual Æquations of the said mean motions of the Sun and Moon, and of the Apogæum and Node of the Moon. The Annual Æquation of the mean motion of the Sun depends upon the Excentricity of the Earth's Orbit about the Sun, which is of $16\frac{1}{2}$ of such parts of which the mean Distance of the Sun from the Earth is 1000; and therefore is call'd the *Æquation of the Center*: And when greatest, it is $1°$. $56'$. $20''$. The greatest Annual Æquation of the Moon's mean motion is $11'$. $49''$. of its Apogæum $20'$. and of the Node $9'$. $30''$. And those four Annual Æquations are proportionable to each other: Therefore when any one of them is the greatest, the three others are the greatest; and when any one is diminish'd the others are also diminish'd in the same proportion: Whence if the Annual Æquation of the Sun's Center be given, you have the other 3 Æquations agreeable to it; therefore the Table of that alone is sufficient. For if the Annual Æquation of the Sun's Center, agreeable to any time taken out of it, be call'd P, and $\frac{1}{10}P = Q$, $Q + \frac{1}{60}Q = R$, $\frac{1}{6}P = D$, $D + \frac{1}{30}D = E$, and $D - \frac{1}{60}D = 2F$; the Annual Æquation of the Moon agreeable to that Time will

Book IV. of Astronomy.

will be R, that of the Moon's Apogæum E, and that of the Node F. It is to be noted, that if the Æquation of the Sun's Center is to be added, the foresaid Æquation of the Moon is to be subtracted, that of the Lunar Apogæum to be added, and the Æquation of the Node to be subtracted: And on the contrary, if the Æquation of the Sun's Center is to be subtracted, the Moon's Æquation is to be added, that of the Lunar Apogæum to be subtracted, and that of the Node to be added.

There is another *Æquation of the Moon's mean Motion*, depending upon the situation of the Moon's Apogæum in respect of the Sun, which is the greatest when the Apogæum of the Moon is in the *Octant* (or at half Right-angles) with the Sun, and none at all when it is come to the Syzygies or Quadratures. This Æquation, when greatest, rises to $3'\,56''$ the Sun being in the Perigæum; but if the Sun be in the Apogæum, not above $3'\,34''$. In other Distances of the Sun from the Earth, this greatest Æquation is reciprocally as the Cube of that Distance. But when the Moon's Apogæum is out of the Octants, the said Æquation becomes less, and is to the greatest Æquation (supposing the distance of the Sun from the Earth to remain the same,) as the Sine of twice the distance of the Moon's Apogæum from the next Syzygy or Quadrature, to the Radius. This is added to the Moon's motion, whilst the Moon's Apogæum passes from the Sun's Quadrature to the Syzygy; but is taken from it, as the Apogæum goes from the Syzygy to the Quadrature.

Moreover, there is another *Æquation of the Moon's motion* depending upon the Aspect of the Nodes of the Moon's Orbit with the Sun;

and it is the greatest when the Nodes are in the Octants with the Sun, and vanishes when they come to the Syzygies or Quadratures. This Æquation is proportional to the Sine of twice the distance of the Node from the next Syzygy or Quadrature, and when greatest rises to 47″. This is added to the Moon's motion, when the Nodes pass from the Sun's Syzygies to his Quadratures; but is substracted, when they pass from the Quadratures to the Syzygies.

From the Sun's true Place take away the mean Motion of the Moon's Apogæum equated as was before shewn; the remainder will be the Annual Argument of the said Apogæum. Then reckon the Moon's Excentricity and the second Æquation of its Apogæum in the following manner,[*which also obtains in the Computation of any other intermediate Æquations:*] Let T (*Fig.* 27.) represent the Earth; TS, a Right line joyning to Earth and Sun; $TACB$ a Right line drawn from the Earth to the mean Place of the Moon's Apogæum equated as before; the Angle STA, the Annual Argument of the said Apogæum; TA the least Excentricity of the Lunar Orbit; TB its greatest Excentricity. Bissect AB in C, and with the Center C, and Interval CA, describe the Circle AFB; let the Angle BCF be drawn equal to twice the annual Argument: TF being drawn will be the Excentricity of the Lunar Orbit; and the Angle BTF the second Æquation of the Lunar Æpogæum. In order to determine them, let the mean distance of the Moon from the Earth, or the Semidiameter of the Lunar Orbit consist of 1000000 Parts; its greatest Excentricity TB, will consist of 66782 of those Parts, and the least TA, or 43319 such Parts; so that the greatest Æquation of its Orbit,

bit, namely when the Apogæum is in the Syzygies, shall be of $7°. 39'. 30''$, or perhaps of $7°\ 40'.\ 00''$; (for there is reason to suspect that that Equation varies according as the Apogæum is in ♋ or ♑;) but when it sticks in the Sun's Quadrature, the said greatest Æquation must be $4°.\ 57'.\ 56''$. and that the greatest Æquation of the Apogæum be of $12°.\ 15'\ 4''$.

Having upon these Principles constructed a Table of the Æquations of the Moon's Apogæum, and the Excentricity of its Orbit for every Degree of the annual Argument, whence the Excentricity TF and the Angle BTF (namely the second and chief Æquation of the Apogæum) agreeable to a given Time, may be easily taken; to the Place of the Moon's Apogæum first equated as before, add the lately found Æquation, if the annual Argument be less than $90°$, or greater than $180°$, but less than $270°$; but if it be otherwise, let it be subtracted from it. The Sum or Difference will be the Place of the Moon's Apogæum equated the second time; which being substracted from the Moon's thirdly equated Place, the remainder will be the Moon's mean Anomaly agreeable to a given Time. Moreover, from that mean Anomaly of the Moon, and the late found Excentricity of the Orb, you will have (by help of the Table of Æquations of the Center of the Moon calculated for every Degree of the mean Anomaly, and some Excentricities, *viz.* 45000, 50000, 55000, 60000, and 65000) the *Prostaphæresis* or Æquation of the Center of the Moon, as commonly; which being subtracted in the first Semicircle or $180°$ of the mean Anomaly, and added in the last, to the Moon's Place, hitherto three times equated, you will have the Moon's Place equated a fourth time.

The Moon's greatest Variation, namely that which happens when the Moon is in the Octants with the Sun, is almost reciprocally as the Cube of the distance of the Sun from the Earth. Let it be taken of 37′. 25″. when the Sun is in the Perigæum, and of 33′. 4″. when it is in the Apogæum: And let the Difference of this Variation in the Octants be reciprocally as the Difference of the Cubes of the Distances of the Sun from the Earth, and from thence construct a Table of the aforesaid variation of the Moon in the Sun's Octants (or of its Logarithm) for every ten, or six, or five Degrees of the mean Anomaly; and for the variation out of the Octants, as the Radius is to the Sine of twice the Distance of the Moon from the next Syzygy or Quadrature; so let the above-found Variation in the Octant be to the Variation agreeable to the given Aspect, which being added to the Moon's Place above-found in the first and third Quadrant (reckoning from the Sun) or subtracted from it in the second and fourth, will give the Moon's Place equated a fifth time.

Again, as the Radius is to the Sine of the Sum of the Distances of the Moon from the Sun and of the Moon's Apogæum to the Sun's Apogæum, (or to the Sine of the excess of that Sum above 360°) so 2′. 10″. to the sixth æquation of the Moon's Place, which is to be subtracted if the foresaid Sum or the said excess be less than a Semicircle, to be added if greater.

As the Radius is to the Sine of the Moon's Distance from the Sun, so also let 2′. 20″. be to a seventh Æquation. Take that away when the

Book IV. *of* ASTRONOMY. 569

the Moon's Light is encreased, and (on the contrary) add it when the Light is diminished; and you'll have the Moon's Place seventhly equated, which is also its Place in its proper Orbit. It is to be observ'd, that the Æquation which is here express'd by the Mean Quantity 2′. 20″. is not always of the same magnitude, but is encreas'd and diminish'd according to the situation of the Lunar Apogæum. For if the Lunar Apogæum be joyn'd with the Sun's Apogæum, the foresaid Æquation becomes greater by 54″; if it be opposite to it, it is as much less: and it varies between the greatest Quantity 3′. 14″; and the least 1′. 26″. And these things hold good when the Lunar Apogæum is in the Sun's Syzygies; but when it sticks in the Sun's Quadrature, the foresaid Æquation is to be diminish'd about 50″ or one whole Minute, when the Moon's Apogæum and the Sun's are in Conjunction; but if they are in Opposition, I can't determine (for want of Observations) whether it is to be encreas'd or diminish'd. Neither can I affirm any thing certain concerning the aforesaid laid down encrease or decrease of the Æquation by 2′. 20″, for want of Observations exact enough.

If the sixth and seventh Æquation are encreas'd or diminish'd in a reciprocal Ratio of the distance of the Moon from the Earth, that is, in a direct Ratio of the Moon's horizontal Parallaxis; they will become more exact. And that may be readily done, if Tables be first constructed for every Minute of the said Parallaxis, and every sixth or fifth Degree as well of the Argument of the sixth Æquation for the sixth Æquation, as of the distance of the Moon from the Sun for the seventh.

From

From the Sun's true Place take away the mean Motion of the Moon's ascending Node equated as before; the remainder will be the Annual Argument of the Node; whence its second Æquation will be computed in the following manner: Let T (*Fig. 27.*) represent the Earth as before; TS a Right-line joyning the Earth and the Sun: Moreover, let $TACB$ represent a Line drawn to the Place of the Moon's ascending Node equated as before, and STA the Annual Argument of the Node. Let TA be taken in the same Ratio to AB as 56 to 3, or $18\frac{2}{3}$ to 1. Bissect BA in C, and with the Center C and Interval CA, or CB, draw the Circle AFB, and let the Angle BCF, be equal to twice the Argument of the Annual Node found as before; and the Angle BTF, will be the second Æquation of the ascending Node, to be added whilst the Node goes from the Sun's Quadrature to the Syzygy, and subtracted as the Node goes from the Syzygy to the Quadrature. And so you have the true Place of the Node of the Moon's Orbit: Whence from Tables construed after the common manner, may be computed the *Moon's Latitude* and the *Reduction of the Moon from its Orbit to the Ecliptic*, supposing the Inclination of the Moon's Orbit to the Plane of the Ecliptic to be of 4°. 59'. 35". when the Nodes are in the Sun's Quadrature; and of 5°. 17' 20". when they are in the Syzygies. From the Longitude and Latitude lately found, and the given Obliquity of the Ecliptic of 23°. 29', the Moon's Right Ascension and Declination will be found.

I put down 57'. 30". for the Parallax of the Moon in the Syzygies, when it is in its mean Distance from the Earth; 33'. 32". 32'''. for the
horary

horary Motion; and 31'. 30". for the apparent Diameter. But in the Moon's Quadratures at a mean distance from the Earth, I make the horizontal Parallax of the Moon to be 59'. 40". the Horary Motion 32'. 12". 2'''. and the apparent Diameter 31'. 3". The mean Distance of the Moon's Center when it is in Octant with the Sun, is distant from the Earth's Center about $60\frac{2}{9}$ Semi-diameters of the Earth.

I set down 10" for the Sun's horizontal Parallax; and 32'. 15". for its apparent Diameter in its mean distance from the Earth.

The Earth's Atmosphere, by refracting and dissipating the Sun's Light, casts a Shadow just as if it was an opake Body, as far as the height of 40 or 50 Geographical Miles: (I call a Geographical Mile the 60th part of a Degree of a great Circle on the Earth's Surface:) This Shadow in an Eclipse of the Moon falling upon the Moon, encreases the Earth's Shadow. And Seconds in the Moon's Disc answers to Miles in the Earth's Atmosphere: Therefore the Semidiameter of the Earth's Shadow projected in the Moon's Disc, is to be encreas'd of about 50 Seconds, (or what comes to the same) in a Lunar Eclipse the horizontal Parallax is to be augmented in a Ratio of about 70 to 69.

If several Places of the Moon nicely observ'd (chiefly about the Quadratures) be compar'd with the Places of it calculated for the same Time in the Theory above; it will appear whether or no there are any sensible Æquations wanting to make it more perfect.

PRO-

V. SIR ISAAC NEWTON'S THEORY OF THE MOON, Lect. 30–31 (pp. 344–368) of William Whiston's *Astronomical Lectures, Read in the Publick Schools at Cambridge* (London, 1728), containing Newton's tract, presented with Whiston's 'perpetual Explication of the Author's Text'

Astronomical Lectures,

Read in the

PUBLICK SCHOOLS

AT

CAMBRIDGE;

By *WILLIAM WHISTON*, M. A.
Mr. *Lucas*'s Professor of the *Mathematicks*
in that UNIVERSITY.

Whereunto is added a

COLLECTION

OF

Astronomical TABLES;

Being those of
Mr. *Flamsteed*, Corrected; Dr. *Halley*;
Monsieur *Cassini*; and Mr. *Street*.

For the Use of young STUDENTS *in the* UNIVERSITY.

And now done into English.

LONDON:
Printed for R. SENEX, in *Salisbury-Court*; and
W. TAYLOR, at the *Ship*, in *Pater-Noster Row*.
MDCCXV.

As Sine 0°. 4'. 4". = = 7. 0729
To Sine 0. 11. 5 = = 7. 5084
So is Tang. 0. 3. 51 = 7. 0491
To Tang. 0. 10. 30 = 7. 4846

From whence it appears, that the Place of *Venus*, observ'd by *Horrox*, both as to Longitude and Latitude, differs very little from the same, as found by the *Caroline* Tables; no more indeed that a few Seconds. And thus much for finding the Places of the Planets by Calculation.

Novemb. 22. 1703.

LECT. XXX.

HAving now Explained the Theory of the Planets, and brought our Astronomical Task to a Conclusion, we should forthwith proceed to the Famous Sir *Isaac Newton*'s Philosophy; but that we are happily detain'd in the present Place for a while by that Celebrated Theory of the Moon, which that Prince of Geometricians and Astronomers hath very lately compos'd, out of Mr. *Flamsteed*'s Observations, and Dr. *Gregory* hath communicated to the Public: It being well worthy to be entertain'd and embrac'd by all Astronomers with the greatest Joy. For it is such a Theory, as when we spake of *Horrox*'s Hypothesis we in vain wish'd for; while we expected it from none but him from whom we now have it. And indeed it is a Theory which the Author hath found, so agreeing with the Heavens, that the Disagreement therefrom, in the Conjunctions with the Sun;

Greg. p. 232.

Astronomical Lectures.

Sun; and this, when greatest, scarce ariseth to more than two, and in the Quadratures scarcely to three Minutes; and for the most Part it is so small, that it seems as if it ought rather to be imputed to the Uncertainty of Observation than any thing else. I shall therefore here present you with this Theory, as I have endeavour'd to illustrate it with a perpetual Explication of the Author's Text.

Sir ISAAC NEWTON's *Theory of the Moon.*

"THE *Greenwich* Observatory is more *West* than that of *Paris* by 2°. 19'. Than *Uraniburg* 12°. 51'. 30". and than that of *Dantzick* by 18°. 48'.

"I suppose the following mean Motions of the Sun and Moon from the Vernal Equinox in the *Greenwich* Meridian; In the Year 1680, the last of *December* in the *Julian* Style, at Noon, the Mean Motion of the Sun was 9s. 20°. 34'. 46". the Motion of the Apogeum of the Sun 3s. 7°. 23'. 30". the Mean Motion of the Moon 6s. 1°. 45'. 45". That of the Apogeum of the Moon 8s. 4°. 28'. 5". That of the Ascending Node of the Lunar Orbit 5s. 24°. 14'. 35". And in the Year 1700, the last of *December Old-Stile*, at Noon, the Mean Motion of the Sun was 9s. 20°. 43'. 50". And that of the Apogee of the Sun 3s. 7°. 44'. 30". The Mean Motion of the Moon 10s. 15°. 19'. 50". that of the Apogee of the Moon 11s. 8°. 18'. 20". and that of the Ascending Node 4s. 27°. 24'. 20". For in 20 *Julian* Years, or 7305 Days, the Motion of the Sun is, 20$^{revolut.}$ 0s. 0d. 9'. 4". the Motion of the Apo-

"gee

"gee of the Sun 21'. 00". The Motion of the
"Moon. 267$^{rev.}$ and 4^s. 13°. 34'. 3". that of the
"Apogee of the Moon 2$^{rev.}$ and 3^s. 3°. 50'. 15".
"The Motion of the Node. 1$^{rev.}$ and 0^s. 26^d. 50.
"15". All the foresaid Motions are from the
"Point of the Vernal Equinox. But if there
"be substracted from these the Motion of the
"Equinoctial Point it self *in antecedentia*, which
"is perform'd in the mean while, to wit, 16'. 40",
"there will remain these Motions in respect of
"the Fixed Stars in 20 *Julian* Years; to wit, the
"Motion of the Sun 19$^{rev.}$ 11^s. 29°. 52'. 24". Of
"the Apogee of the Sun 4'. 20". The Motion of
"the Moon 267$^{rev.}$ 4^s. 13°. 17'. 25". that of the
"Apogee of the Moon, 2$^{rev.}$ 3^s. 3°. 33'. 35". that
"of the Node of the Moon, 1$^{rev.}$ 0^s. 27°. 6'. 55".
"According to this Computation, the Tropical
"Year consists of 365^d. 5^h. 48'. 57". and the Sy-
"dereal Year of 365. 6. 9. 14½." Thus far
Sir *Isaac Newton*. All which is easy enough to
to be understood, if these following Things be
well minded.

(1.) That we have here two Epocha's or Roots
of the Mean Motions of the Sun and Moon.
Whereof one is to be applied to *A. D.* 1680. the
other to the Year 1700, but both to the End of
the Year, and the Mean or Middle Time.

(2.) That here we have a Table of the compleat Mean Motions of Twenty Years most accurately compos'd; from whence a Table of Roots,
and another of compleat Years may be without
difficulty fram'd for any Number of Years,
Months, Days, *&c.* whether past or to come.

(3.) That here a Motion of the Sun's Apogee is suppos'd. Which is to be wonder'd at;
when

when * this Author himself elsewhere hath taken all Motion from the Aphelia, as well as from the Nodes of the primary Planets; save some very small Mutations not to be regarded; yea, and hath superstracted the Immobility of the Fixed Stars upon the Immobility of the same; And when Mr. *Flamsteed* himself, to whose Observations and Hypotheses our Author hath in some sort tied himself in the present Theory, doth with *Street* suppose and assert the absolute Rest of the Earth's Aphelion. It is therefore to be wonder'd at, that after that this Motion had been exploded out of Astronomy, to the great Satisfaction of the Learned, these Great Men should think that it ought to be brought in again. But be this as it will, the Motion here suppos'd is so small, that it scarce makes one Degree in 300 Years space; and consequently that for many Ages on this side and on that, it matters not much whether we suppose the Aphelion to be mov'd or to be unmoveable.

(4.) That all the rest of the Motions, besides that of the Nodes, is *in Consequentia*, or from *West* to *East*; but that of the Nodes the other way, or from *East* to *West*. When therefore the Author bids that we substract from the former Motions, the Motion of the Equinoctial Point *in Antecedentia*, that the said Motions, consider'd in respect of the Fixed Stars, may be found: this Substraction belongs only to the rest of the Motions, and not to that of the Nodes; for this is not to be substracted, but added, as the Author's Numbers themselves do shew. But he goes on.

---

* Phil. Nat. Prin. Math. l. 3. prop. 14.

" The

"The Mean Motions of the Luminaries above
"suppos'd, are affected with many Inequalities:
"And first, there are the Annual Equations of
"the said Mean Motions of the Sun and Moon,
"and of the Apogee and Nodes of the Moon:
"The Annual Equation of the Mean Motion
"of the Sun depends upon the Eccentricity of
"the Orbit of the Earth about the Sun, which
"consists of $16\frac{11}{12}$ such Parts as the Mean Di-
"stance of the Sun from the Earth contains
"1000; and from thence it is called the Equa-
"tion of the Center; and it is when greatest
"$1^d. 56'. 20''$. The greatest Annual Equation of
"the Mean Motion of the Moon is $11'. 49''$.
"that of the Apogee of the same $20'$. and that
"of the Node $9'. 30''$. And these four Annual
"Equations are always proportional amongst
"themselves; and therefore when any of them
"is the greatest, the rest are also each of them
"the greatest; and any one of them being di-
"minish'd, the rest are diminish'd in the same
"proportion. From whence the Annual Equa-
"tion of the Center of the Sun being given,
"there are also given the Three other Equati-
"ons. And therefore a Table of that mention'd
"Equation sufficeth for all. For if the Annual
"Equation of the Center of the Sun agreeing
"to any Time, which is taken from the Table,
"be called P; and $\frac{1}{10}$ P be put $=$ Q. and Q$+\frac{1}{60}$
"Q $=$ R. and $\frac{1}{6}$ P $=$ D. D$+\frac{1}{30}$ D$=$E. and
"D$-\frac{1}{60}$ D $=$ 2 F. The Annual Equation of
"the Moon agreeing to the same Time, will be
"R; the Equation of the Moon's Apogeum E,
"and of the Node F. It is to be noted, that if
"the Equation of the Center of the Sun is to be
"added, the foresaid Equation of the Moon is
"to

Astronomical Lectures. 349

" to be substracted; the Equation of the Moon's
" Apogeum is to be added; but the Equation of
" the Node substracted. And on the other hand,
" If the Equation of the Center of the Sun is to
" be substracted, the Equation of the Moon is
" to be substracted; That of the Apogeum to be
" substracted; but that of the Node to be added."
Thus far Sir *Isaac Newton*.

Our Famous Author in this Place treats of four Kinds of Equations, depending on the Anomaly of the Sun or Earth. The first is that most known one, whereby we gather the Earth's Coequate out of the Mean Anomaly; which indeed must be proportionale to the Eccentricity of the Earth's Orbit; and this Eccentricity he defines to consist of $16\frac{11}{12}$ such Parts, as the Mean Distance of the Sun from the Earth contains 1000 of them. And this he determines from Mr. *Flamsteed*'s Observations. For A. D. 1692, Mr. *Flamsteed* observ'd, as he sometime wrote to me, that the Eccentricity of the Earth's Orbit consists of 1692 such Parts as the Mean Distance of the Sun contains 100000; which is plainly the same Proportion as that here set down, as any one that tries will find. And from this Eccentricity our Author gathers, that the greatest Quantity of Equation, (which is to be applied about the Middle of the Ellipsis) is $1^d. 56'. 20''$. which Equation is much less than the *Caroline*. The Second Equation is that of the Middle Motion of the Moon, which ought to be mended and corrected according to the various Distance of the Sun from the System of the Earth and Moon. For the Middle Motion of the Moon, which was formerly determin'd, is only a Mean Motion in a more lax Sense; to wit, as agreeing to this Planet, consider'd as in

the

the Space of many Years together. For as we noted heretofore, The Moon's Periodical Motion is the shorter when the Earth is in the Aphelion, and the Sun is the most Remote; but as the Earth departs from thence, and approaches to the Sun, the Periodical Time is increas'd; until the Earth being now arriv'd to the Perihelion, it becomes the greatest of all. Therefore the Middle Place of the Moon before defin'd, is to be corrected by the Addition or Substraction of a certain Arch or Angle which depends upon the Earth's Mean Anomaly, or rather upon the Sun's Distance, in great Measure proportionale to the same: And this Equation, when it is the greatest, our Author determines to be but 11'. 49" about the Mean Distance of the Sun. The Third Equation is of the Apogee of the Lunar Orbit; the Motion of which, seeing it altogether depends upon the Sun, as we have noted before; It is no wonder that the same is lesser according to the greater Distance of the Sun, and greater according to the less Distance of the same. And this Equation, when it is the greatest, he defines to be of 20' only. The Fourth Equation is of the Node of the Lunar Orbit; and seeing the Regress or Motion thereof *in Antecedentia*, doth likewise arise from the Sun, as will be demonstrated in its Place, it is reasonable that we expect to find the same lesser in the greater Distance, and greater in the less Distance of the Sun; and the Quantity thereof is here determin'd to be 9'. 30". All these Equations therefore are to be estimated according to the greater or lesser Distance of the Sun, or according to the Sun's Anomaly, and they are proportional one to another, or have always the same proportion betwixt themselves

which

Astronomical Lectures. 351

which the Numbers have that are set down above, as expressing the Quantity of every one; to wit, as being amongst themselves in proportion, as $6980'' : 709'' :: 1200'' : 570''$: or in our Author's Phrase; If the Equation of the Center be called P, the Equation of the Middle Motion of the Moon will be every-where $\frac{61}{600}$ P; the Equation of the Apogeum of the Moon $\frac{31}{180}$ P; and the Equation of the Node $\frac{29\frac{1}{2}}{260}$ P. So that by a Table of the Equation of the Center, all these Equations may every-where be found; and there needs not a particular Table for every one. But forasmuch as when the Sun is most remote, both the Motion of the Earth, and those of the Lunar Apogee and Node are the slowest; and are each of them the swiftest, when the Sun is the nearest: It is manifest that all these Equations are of the same sort, that is, are at the same time to be added, and at the same time to be substracted. Whereas on the contrary, when the Sun is most remote, the Periodical Middle Motion of the Moon is the swiftest, and the slowest of all when the Sun is the nearest: So that this Equation is of a diverse Sort or Title from the rest, and is to be added when they are substracted, and so on the contrary. And this is the plain meaning of the Author's Canon. For although he asserts that the Equation of the Node is to be substracted, when the Equation of the Center of the Sun is added, and on the contrary; yet seeing he doth intend that that Equation, together with the rest, is to be added according to the Series

ries of the Signs, he means nothing else but a Substraction from the true Direction of the Regress of the Nodes, and so on the contrary. But he goes forward.

"There is another Equation of the Mean Motion of the Moon, depending upon the Situation of the Lunar Apogee with respect to the Sun; which is greatest of all when the Moon is an Octant with the Sun; and none at all when it comes to the Conjunctions; [Conjunctions I say in the Plural, for under that Word I here include the Opposition also] and Quadratures. This Equation, when it is greatest, ariseth to 3′. 56″, the Sun being in the Perigee; but if it be in the Apogee, it exceeds not 3′. 34″. In other Distances of the Sun from the Earth, this Equation is greatest, as is the Cube of that Distance reciprocally: But when the Moon's Apogee is without the Octants, the said Equation becomes lesser; and it is to the greatest, for the same Distance of the Earth and Sun, as the Sine of the double Distance of the Lunar Apogee from the next Conjunction or Quadrature, is to the Radius. This is added to the Motion of the Moon, whilst the Apogee of the Moon passes from the Square of the Sun to a Conjunction; but is substracted from thence in the Transit of the Apogee from a Conjunction to a Quadrature."

In which Words our Author explains a Second Equation of the Middle Motions of the Moon; that, to wit, which depends upon the divers Position of the Lunar Apogee. For hitherto we have only attempted to equate the Mean Motion of the Moon as perform'd in a Circle, or Mean Ellipsis: But seeing the Ellipsis in which the Moon

is

is mov'd is very various and inconstant, and the Eccentricity thereof is varied above a 3d Part; so that when the Apsides of the Moon are in the Conjunctions, the Eccentricity of the Orbit is increas'd; and when they are in the Quadratures it is diminish'd, and the Ellipsis comes nearer to a Circle; It is requisite that we should reckon the Moon's Periodical Time, as increas'd in the former Case, because of the increase of the Area, or at least the Circumference of the Orbit, and diminish'd in the latter Case, because of the Diminution of that Area, or at least the Circumference of the Orbit: From whence a 2d Equation is requir'd, that we may have the Middle Motion equated also according to the various Position of the Apogee: But that this Equation is sometimes greater, sometimes lesser, according to the Reciprocal Proportion of the Cubes of the Sun's Distance, is what will be demonstrated when we undertake the Philosophy of this Famous Author; and we shall in the mean while pass it over, as not properly belonging to this Place, where we treat of Matters purely Astronomical. But as for that Assertion, that every intermediate Quantity of Equation is as the Sine of the double Distance of the Lunar Apogee from the next Conjunction or Quadrature; this is what is perpetual and necessary, wheresoever any Motion which is to be corrected by Equation, doth in the entire Revolution come twice unto the greatest, and twice unto the least Quantity: and this in such sort, that a Quadrant in the Heavens should answer to a Semi-circle in Calculation; and therefore it is absolutely necessary that a Degree in the Semi-circle should answer to half a Degree in the Quadrant, and so

A a perpe-

perpetually. The Rest of the Things in the present Section need no further Explication. Sir *Isaac Newton* proceeds thus. "Moreover there is
" another Equation of the Motion of the Moon,
" which depends upon the Aspect of the Nodes
" of the Lunar Orbit with respect to the Sun;
" and it is greatest when the Nodes are in the
" Octants of the Sun; and vanisheth away wholly
" when they arrive at the Conjunctions or Qua-
" dratures. This Equation is proportionate to
" the Sine of the double Distance of the Node
" from the next Conjunction or Quadrature; and
" when it is greatest amounts to 47″. This is
" added to the Motion of the Moon, whilst the
" Nodes pass from the Sun's Conjunctions to the
" Quadratures of the same; and is substracted in
" their Transit from the Quadratures to the Con-
" junctions.

In which Words our Author sets forth a Third Equation of the Mean Motions of the Moon; which indeed is very small, but yet not altogether to be neglected. It depends upon the Position of the Nodes of the Lunar Orbit with respect to the Sun; for hitherto we have endeavour'd to equate the Mean Motion of the Moon as perform'd in the same Plane. But seeing the Plane of this Orbit is continually changing, except when the Nodes are posited in the Conjunctions, at which time they wholly rest; It is manifest that at that time only the Moon possesseth the same Plane the Month throughout; and consequently describes the least Area in the same Plane, and therefore performs her Periods in the least time: Whereas at other times a greater Area is describ'd by reason of a great many Planes as it were, which are its way, and that really drawn

out

Astronomical Lectures.

out into a Curve Superficies; and especially when the Nodes are about the Quadratures, for then there is the greatest Variety of Planes, and consequently the Periodical Time must necessarily be the greater. When therefore the Nodes are in the Conjunctions, the Periodical Motion of the Moon is a little swifter, and when they are in the Quadratures a little slower; so that it is not to be wonder'd at that the Equation in the former Case is Addititious, in the latter Ablatitious, as our Author's Canon requires. But it is to be observed here and every-where, that as well the Nodes as the Apogee of the Moon do seem in respect of the Sun to be mov'd from *East* to *West*; and that perpetually; to wit, because of the apparent Motion of the Sun from *West* to *East*. For although the Apogee be carried to the same Part with the Sun, and the Nodes to the contrary Part, each according to its proper Motion; yet these Motions, if compared with the Annual Motion of the Sun, are so small, that in this respect both of them ought to be neglected, as if they were no Motion at all; and consequently the Mutations of which we now speak are to be estimated from the Motion of the Sun; which while it tends to the *East*, makes that the two Motions aforesaid should seem to tend to the *West*. Which things have been spoken for preventing Mistakes, which we may easily fall into in these Matters. But whereas the present Equation is said to be proportional to the Sine of the double Distance of the Node from the next Conjunction or Quadrature, this is what we met with in the former Section, and is derived from the same Causes; and so I need to say no more concerning it in this Place. Our Author adds.

"Subtract from the True Place of the Sun the
"Middle Motion of the Apogee of the Moon,
"so equated as was shew'd before; the Remainder
"will be the Annual Argument of the said Apo-
"gee. From thence let there be computed the
"Moon's Eccentricity and the Second Equation
"of the Apogee thereof in the following man-
"ner," [which, as Dr. *Gregory* observes, hath place also in the Computation of all the Inter-mediate Equations whatever.]

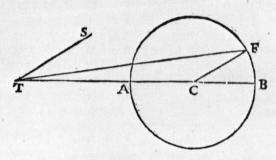

"T represents the Earth: TS a Right Line
"joining the Earth and the Sun. TACB a Right
"Line drawn from the Earth to the Middle
"Place of the Lunar Apogee, equated as above;
"The Angle STA is the Annual Argument of
"the said Apogee. TA the least Eccentricity of
"the Lunar Orbit. TB the greatest Eccentricity
"of the same. Bisect AB in C. From the Cen-
"ter C through A describe the Circle AFB. Let
"the Angle BCF be made double to the Annual
"Argument; the Right Line TF, when drawn,
"will be the Eccentricity of the Lunar Orbit;
"and the Angle BTF will be the Second Equa-
"tion of the Moon's Apogee. For the Deter-
"mination of these things, let the mean Distance
"of

"of the Moon from the Earth, or the Semi-dia-
"meter of the Lunar Orbit, confift of 1,000,000
"Parts. Its greateft Eccentricity TB will be of
"66,782; and the leaft Eccentricity TA 43,319
"of the fame Parts. So that the greateft Equa-
"tion of the Orb thereof, to wit, when the Apo-
"gee is in the Conjunctions, is $7°. 39'. 30''$, or
"perhaps $7°. 40'$; (for it is fufpected that this is
"chang'd according to the Situation of the Apo-
"gee in *Cancer* or *Capricorn*); but when the Apo-
"gee is in the Quadratures of the Sun, then the
"faid greateft Equation is $4^{gr}. 57'. 56''$. From
"whence confequently the greateft Equation of
"the Apogee is, $12°. 15'. 4''$. Which Words of
our Author are fo eafy to be underftood, that
they require not our Explication.

Novemb. 29. 1703.

LECT. XXXI.

OUR Celebrated Author proceeds. "There
"being fram'd according to thefe Princi-
"ples a Table of the Equations of the Apogee
"of the Moon, and of the Eccentricities of the
"Orbit of the fame for each Degree of the
"Annual Argument; from whence the Ec-
"centricity TF and the Angle BTF, (to wit,
"the Second and chief Equation of the A-
"pogee), convenient for the Given Time,
"may eafily be fetch'd, Unto the Place of
"the Apogee of the Moon firft equated, as
"above, let there be added the Equation juft
"now found, if the Annual Argument be lefs
"than 90 Deg. or greater than 180 Deg. but
"lefs than 270. Otherwife let the faid Equation
"be

" be substracted: The Sum or Difference will be
" the Place of the Moon's Apogee equated a 2d
" Time: Which being substracted out of the
" Moon's Place the 3d Time equated, there will
" be left the Mean Anomaly of the Moon, a-
" greeing to the Given Time. Moreover, from
" the Moon's Mean Anomaly and the now found
" Eccentricity of the Orbit, (by means of a Ta-
" ble of Equations of the Center of the Moon
" framed for each Degree of the Mean Anoma-
" ly, and for several Eccentricities, as 45000,
" 50000, 55000, 60000, and 65000,) there will
" be had the Prosthaphæresis or Equation of the
" Center of the Moon, as in the common way;
" which being substracted in the former Semi-
" circle of the Mean Anomaly; but added in the
" latter, to the Place of the Moon, the 3d Time
" equated, there will come forth the Moon's Place
" the 4th Time equated". Thus far our Author.

For the right Understanding of which, we are
to observe, that the Equation of the Apogee, or
Prostaphæresis is to be deduc'd out of such a Tri-
gonometrical Calculation as follows. In the Triangle TCF there are given the Side TC of 55050½
Parts, and that CF of 11731½, and the intercept-
ed Angle TCF; to wit, the Complement of the
given Angle FCB to two right ones: From whence,
by the known Method, will be found the Angle
CTF, the Measure of the Equation sought: For
as the Sum of the Sides TC and CF=TB, which
is of 66782 Parts, is to their Difference TA of
43319 Parts (both which Quantities are every-
where fixed and the same;) so is the Tangent of
the Half-Sum of the Angles CTF, and CFT=to
half the given Angle FCB, to the Half-differ-
ence of the same; which being taken away from
that Half-Sum, will give the Angle CTF, or the
sought

sought Equation of the Apogee. And as for the Eccentricity TF, that will easily be known out o what has been already found: For in the same Triangle TCF, whose Angles are already given, the Side TF will become known in this manner: As the Sine of the given Angle CTF is to the opposite given Side CF; so likewise is the Sine of the given Angle TCF, or which comes to the same, the given Angle FCB, to the Side TF. Q.E.I

But our Author goes on thus. "The greatest Variation of the Moon, (to wit, that which happens when this Planet is in the Octants of the Sun,) is well nigh reciprocally, as the Cube of the Distance of the Sun from the Earth. Let it be taken to be 37′. 25″, when the Sun is in the Perigee; and 33′. 4″, when in the Apogee; and let the Difference of this Variation in the Octants be made reciprocally, as the difference of the Cubes of the Distances of the Sun from the Earth; and from thence let there be fram'd a Table of the foresaid Variations of the Moon in the Octants of the Sun (or of the Logarithms thereof) for each ten, or six, or five Degrees of the Mean Anomaly; and for the Variation without the Octants, let it be made thus; As the Radius, is to the Sine of the double Distance of the Moon from the next Conjunction or Quadrature; so is the above-found Variation in the Octant, to the Variation agreeing to the given Aspect: which being added to the Place of the Moon above found in the first and third Quarter (reckoning from the Sun;) or substracted from the same in the 2d or 4th Quarter, will give the Moon's Place equated the 5th Time." Which Things are so clearly propounded by our famous Author, that they need little Explication. We therefore shall only shew

our Reader the way in an Example or two. Let the greatest Variation be to be found, where the Sun's Distance is in a Mean, or of 1000 Parts. The Cube of the greatest Distance, which is of about 1017 Parts, is 1,041,871,913; where the Variation by the Hypothesis is 33′. 4″. The Cube of the least Distance, which is of about 983 Parts, is 949,862,087. The Difference of the Cubes, of the greatest and least, is 92,009,826. From whence the Analogy is thus Ordered: As that Difference of the Cubes 92,009,826 is to the Difference of the Cubes of the Mean Distance 1000, and the least 983, to wit, 50,137,913; so is the Difference of the greatest and the least Variation 4′. 21″, or 261″ to a 4th Number 142″, or 2′. 22″, to be superadded to the former least Quantity: Or as the Difference of the greatest and least Cubes 92,009,826, is to the Difference of the Cubes of the Mean Distance 1000, and the greatest 1017, to wit, 41,871,913; so is the Difference of the greatest and least Variation, or 261″ to a 4th Number 119″, or 1′. 59″ to be taken away from the greatest foregoing Quantity. For by both ways the same Variation, agreeing to the Sun's Mean Distance, will come out, to wit 35′. 26″. But it is to be noted, that there is no need of several Figures of the former Terms of the Analogy; which will make the Work much shorter; for as 92 : 50 :: 261″ : 142″. And as 92 : 42 :: 261″. 119″. which it is oftentimes worth the while to observe in this and the like Cases. But now let us hear Sir *Isaac Newton*, who saith;
" Again, as the Radius, is to the Sine of the Sum
" of the Distances of the Moon from the Sun,
" and of the Apogee of the Moon from the A-
" pogee of the Sun (or to the Sine of the Ex-
" cess of that Sum above 360^d;) so is 2′. 10″, to
" the

Astronomical Lectures. 361

" the 6th Equation of the Place of the Moon;
" to be subſtracted if the foreſaid Sum, or the
" ſaid Exceſs be leſs than a Semi-circle; to be ad-
" ded if greater.

How it ſhould come to paſs that this 6th E-
quation of the Moon ſhould ariſe from Cauſes
which are ſo unlike join'd together amongſt
themſelves, as are the Motion of the Moon from
the Sun, and the Motion of the Apogee of the
Moon from the Apogee of the Sun; I muſt acknowledge my ſelf to be altogether Ignorant; nor
is there Opportunity for enquiring in theſe Matters merely Aſtronomical. In the mean while,
I ſuſpect that this Equation was rather deduc'd
from Mr. *Flamſteed*'s Obſervations, than from Sir
Iſaac Newton's own Argumentation; otherwiſe we
had had two Equations aſſign'd each to its proper Cauſe out of this conjunct One. However, the
Words of our Author are too clear to need that
we ſhould add one Word for interpreting them.
He proceeds. " Let it be made, as the Radius,
" is to the Sine of the Diſtance of the Moon
" from the Sun; ſo is 2'. 20" to the 7th Equati-
" on; take away this when the Moon's Light is
" increas'd, and add it when it is diminiſh'd, and
" there will come forth the Moon's Place equa-
" ted the 7th Time; which is the Place thereof
" in its proper Orbit. It is to be noted that this
" Equation which here is expreſs'd in a Mean by
" 2'. 20", is not always of the ſame Quantity,
" but is increas'd and diminiſh'd according to the
" Situation of the Moon's Apogee: For if the
" Apogee of the Moon be in Conjunction with
" the Apogee of the Sun, the foreſaid Equation
" is greater by about 54"; but if it be in Oppo-
" ſition to the ſame, it is ſo much leſs: and it is
" betwixt the greateſt Quantity 3'. 14", and the
" leaſt

"leaft 1'. 26". And this holds when the Apo-
"gee of the Moon is in the Conjunctions with
"the Sun; but when it is in the Quadratures,
"the foresaid Equation is to be diminish'd by
"about 50", or one whole Minute, when the
"Apogee of the Moon, and that of the Sun, are
"in Conjunction; but if they be in Opposition,
"by reason of the Want of Observations, I can-
"not affirm whether it be to be increas'd or di-
"minish'd; yea, I dare not certainly determine
"of the above-suppos'd Increase or Decrease of
"the foresaid Equation 2'. 20", from the defect
"we are in of Observations, sufficiently exact.
"If the 6th and 7th Equations be increas'd or
"diminish'd in the Reciprocal Proportion of the
"Distance of the Moon from the Earth, that is,
"in the Direct Proportion of the Moon's Hori-
"zontal Parallax, they will become the more
"Accurate. And this will readily be done, if
"first of all Tables be fram'd for the Moon's
"Horizontal Parallax, and each six or five De-
"grees thereof, together with the Argument of
"the 6th Equation for the 6th Equation, and
"then that of the Distance of the Moon from
"the Sun for the 7th." Thus far our Author.

Now for the better understanding this Equation, the 7th and last of those which respect the
Longitude of the Moon in her own Orbit, we
must know, that the Earth possesses the A-
rithmetical Center, as it were, of the Moon's
Orbit, not the Geometrical; that is, that it
possesses Arithmetically, but not Geometrically the Middle Point, betwixt the least Distance of the Moon from the Sun, which is in
the New-Moon, and that at the Full-Moon,
which is the greatest; for the Distance of the
Moon from the Earth in general is the same in
the

Astronomical Lectures. 363

the Full-Moon, as in the New-Moon *cæteris paribus*; when nevertheless that same Distance hath a less proportion to the Distance of the Sun from the Moon at the Full-Moon, which is then greater, than it hath at the New-Moon, at which Time the said Distance is less, as is manifest to every one. From whence it comes that a Mean Geometrically Proportional betwixt the Distances of the Moon from the Sun, which is greatest at the Full-Moon, and least at the New-Moon, will always be terminated betwixt the Sun and the Earth; and that intermediate Point will constitute a certain Center of the Lunar Motions, something distant from the Center of the Earth, and towards the Sun. And if the Force of the Sun, by which the Motion of the Moon is disturb'd, were only in a Reciprocal Proportion of the Distance, that Point would be a certain Center of the Lunar Motions and Equations, about which, as about the Center of more even Motions, [rather than that which hath been hitherto suppos'd] the Astronomical Computation would be rightly ordered; so likewise if the Force of the Sun, which disturbs the Lunar Motions, were in a Duplicate Reciprocal Proportion of the Distances, there would be some middle Point betwixt the Sun and the Earth, and a little more remote from the Center of the Earth, which might rather be nam'd the Center of the Lunar Motions than the middle Point of the Earth it self; And the same thing would happen much more forcibly, were it suppos'd to be in the Triplicate Reciprocal Proportion of the Distances increas'd or diminish'd; and so forwards.

Since therefore we have hitherto suppos'd the Equations in the New-Moons and Full-Moons; in the Octants towards the Sun, and those towards

wards the opposite Part, and all other Mutations of Motions to be equal; which yet could not hold good, except about a certain Center placed between the Center of the Earth and the Sun, though near to the Center of the Earth; It is therefore reasonable that we reduce all things from that Center unto the Center of the Earth: Which will be done in the same manner, as by the Common Prosthaphæreses we reduce unto a certain Equality the Motion of the Planets, which is, almost, Even about the Superior Focus of the Ellipsis, but Uneven about the Sun, which is placed in the Inferior Focus; and determine for any given Moment of Time, and this with respect to the Center of the Sun, where a Planet is placed in its own Orbit. For Illustration's sake, let us suppose the Orbit of the Moon to be something Eccentrical; that the Line of the Apsides, and that of the Apogee, does respect the Center of the Sun; and let us reckon that Center of intermediate Equations to be, as it were, the Superior Focus, and the Center of the Earth, as it were, the Inferior Focus; and determine the Eccentricity of the Orbit to be so very little, that it yields an Equation of no more than 2'. 20". what follows hence will make the present Computation easy enough. Excepting that the Changes of that Quantity 2'. 20", according to the various Positions of the Apsides and Conjunctions, cannot be deduc'd by that way of Reasoning. Nor is this to be wonder'd at, since Sir *Isaac Newton* himself seems to doubt of them, and will not venture to propound the same to his Reader any otherwise than from Observations. This Equation certainly is subject to Mutation, and that by reason of very many Causes, which indeed to deduce severally and distinctly would

be

be a most difficult Task. But our Business at present is with Things purely Astronomical, and therefore we will in no wise make any Doubt about the Equations, or Quantities and Changes of Equations mention'd by Mr. *Flamsteed* and Sir *Isaac Newton*; Into the Causes of all which we shall have hereafter a more proper Place to enquire. But as for what our Author adds concerning the Increase and Decrease of the 6th and 7th Equation in the Direct Proportion of the Moon's Horizontal Parallax, that I suppose means, that the same Equations are adapted to the mean Parallax or Distance of the Moon; and that when the Parallax ariseth above Mean, the Equations are to be increas'd; and when it is below the Mean, to be diminish'd. But our Author proceeds thus. " Substract from the
" True Place of the Sun, the Middle Motion of
" the Ascending Node of the Moon, equated as
" above; the Remainder will be the Annual Argument of the Equation; from whence the
" Second Equation of the same will be equated
" in the following manner. In the foregoing
" Figure (Page 356.) let T, as before, represent the Earth; TS a Right Line joining the
" Earth and the Sun. Moreover, let TACB represent a Right Line drawn to the Place of
" the Ascending Node of the Moon, equated as
" above; and let STA be the Annual Argument
" of the Node. Let TA, taken to AB, as 56
" to 3, or as $18\frac{2}{3}$ to 1. Bisect BA in C, and from
" the Center C, with the Interval CA or CB
" describe the Circle AFB: and let the Angle
" BCF be made equal to the double Annual Argument of the Node, as found above. And
" the Angle BTF will be the Second Equation
" of the Ascending Node, to be added in the
" Transit

"Transit of the Node from the Conjunctions to
"the Quadratures, but to be substracted in the
"Transit of the same from Quadratures to the
"Conjunctions: and thus we obtain the True
"Place of the Node of the Lunar Orbit. From
"whence by the help of Tables fram'd accord-
"idg to the Common Method, the Moon's La-
"titude will be computed, and the Reduction
"of the Moon from her own Orbit to the E-
"cliptic; the Inclination of the Lunar Orbit
"to the Plane of the Ecliptic being suppos'd
"$4^{gr}. 59'. 35''$, when the Nodes are in the Qua-
"dratures of the Sun, and $5^{gr}. 17'. 20''$ when
"they are in the Syrygies. And then from the
"now found Longitude and Latitude, and the
"given Obliquity of the Ecliptic $23^{deg}. 29'$, the
"Right Ascension and Declination of the Moon
"will be collected". Which Words of our Author will be easily enough understood if we note
these Things also. (1.) That for this Reason a-
lone TA is to be taken to be AB, as 56 to 3;
to wit, because from thence the greatest Angle
FTC will become equal to the greatest E-
quation of the Node given by Observation.
(2.) That the particular Inclination of the Lunar Orbit to the Ecliptic is every Time to be
defin'd by this Analogy; namely, As the Difference of the Right Sines of the greatest Inclination, which is $5°. 17'. 20''$, and of the least,
which is $4°. 59'. 35''$; so will the Right Sine of the
Distance of the Lunar Node from the next Conjunction be to a 4th Number; which Number being superadded to the Sine of the least Inclination, or taken away from the Sine of the greatest, will give the Sine of the Inclination sought.
3. The Longitude and Latitude being given, the
Moon's Right Ascension and Declination will be
given

Astronomical Lectures. 367

given in this manner. Let a great Circle of the Sphere be understood to be drawn from the Point of the Vernal Equinox to the Center of the Moon; there will be a Spheric Oblique-angled Triangle; by the Resolution of which, almost all the Difficulty of the Problem will be taken away; for there will be given one Side, to wit, the Distance betwixt the Point of the Vernal Equinox and the Lunar Node; and another Side also will be given, to wit, the Distance of the Moon from the Node; and there will likewise be given the inclosed Angle, as being the Complement of the Angle of the Inclination of the Lunar Orbit to the Ecliptic. From whence, by the Famous *Caswell*'s 9th Case of Spheric Oblique-angled Triangles, the Hypotenuse will be given; and by the 10th Case, the Angle adjacent to a Side that toucheth the Point of the Vernal Equinox; which Angle so found being added to the Angle of the Inclination of the Ecliptic to the Equator, which here is most accurately defin'd to be of 23°. 29′, or taken away from the same, according to the Various Position of the Lunar Nodes, will give the Angle at the Equator. From which and the Hypotenuse already given, the Moon's Right Ascension and Declination from the Equator will easily become known. And no more needs to be said in so plain a Matter. But our Famous Author proceeds thus: " I suppose the Horizon-
" tal Parallax of the Moon in the Conjunctions,
" and in a mean Distance from the Earth, to be
" 57′.30″; the Horary Motion thereof 33′.32″.32‴;
" and the Apparent Diameter 31′.30″. But the
" Parallax of the same in the Quadratures at a
" mean Distance from the Earth likewise 56′.40″;
" the Horary Motion 32′. 12″. 2‴, and the Ap-
" parent Diameter 31′. 3″. The Center of the
" Moon

"Moon in an Octant of the Sun, and meanly distant, is distant from the Center of the Earth about 60⅓ Semi-diameters of the same. I suppose also the Sun's Horizontal Parallax to be 10″, and 32′. 15″ to be the Measure of the Apparent Diameter thereof in a middle Distance from the Earth. The Atmosphere of the Earth, by Refracting and Dissipating the Light of the Sun, casts a Shadow as though it were Opake, unto the Height at least of 40 or 50 Geographic Miles; (by a Geographic Mile, I mean the Sixtieth Part of a Degree of a great Circle upon the Surface of the Earth:) This Shadow in a Lunar Eclipse falling upon the Moon, makes the Shadow of the Earth greater than otherwise it would be, and to each Mile of the Terrestrial Atmosphere there answer one 2d Minute in the Discus of the Moon: And consequently the Semi-diameter of the Terrestrial Shadow on the Discus of the Moon is increas'd by about 50 Seconds; or which comes to the same thing in an Eclipse of the Moon, the Horizontal Parallax of the Moon is increas'd in the Proportion of 70 to 69, or thereabouts."

Which Words of our Author are easy enough in themselves; if we consider that the Semi-diameter of the Lunar Orbit contains about 60 Semi-diamters of the Earth, in the mean while that one first Minute, which answers here to one Mile, contains also 60 Seconds; and if withal we remember that 50 Geographic Miles are about a 70th Part of the Semi-diameter of the Earth. But here I shall conclude. For we have now at length finish'd our Astronomy; and shall defer the Explication of Sir *Isaac Newton*'s Astronomy and Philosophy until the next Term.

Decemb. 6. 1703.

F I N I S.